Norman Jacobs & Kerry O'Quinn present

ROBOTS

Researched & written by Robert M. Hefley
with Howard Zimmerman

Editor: Howard Zimmerman
Design Director: Robert P. Ericksen
Designer: Robert Sefcik

Associate Designer: Elaine Ashburn-Silver
Art Assistant: Laura O'Brien
Contributors: Bob Woods
David Hirsch
David Hutchison

A STARLOG PRESS BOOK

ABOUT THE COVER:

Front Cover, (Clockwise from top left): C3PO and R2D2 getting into trouble in *Star Wars*; Yul Brynner—as an android—gets a facelift in *Westworld*; Gort stands ready to aid Klaatu in *The Day The Earth Stood Still*; guardian alien android being formed in *The Questor Tapes*; Huey, Dewey & Louie, the three droids from *Silent Running*; Cylon warrior from *Battlestar Galactica*; Robby the Robot goes wild in *The Invisible Boy*.

Back Cover, (Clockwise from top left): Robot Twiki and clockwork computer Dr. Theopolis, from *Buck Rogers*; robot daggit Muffit II, from *Galactica*; scientist attacked by *GOG*; masked humans hide their emotions from their robot servants in "One Moment of Humanity," from *Space: 1999*.

STARLOG Press
O'Quinn Studios, Inc.
Norman Jacobs / Kerry O'Quinn
475 Park Avenue South
New York, NY 10016

0-931064-12-0 FIRST EDITION

PREFACE

Biblical legends tell us that God created Man in His own image. This is a supremely egocentric process and one that the human race has desired to duplicate with our *own* creations.

When we became technically advanced and were able to build working robots, they were much more specialized in their functions than the versatile human body—as a result, our robots did not end up looking in the least humanoid.

In fact, most electro-mechanical robots look more like manufacturing assembly lines than *people*. What a disappointment.

Fortunately, science fiction (like all arts) deals more with the *spirit* of things than with the everyday realities. Science fiction writers, artists, and filmmakers have given us what science has not found practical—machines fashioned in our own image.

There is a wonderous sense of achievement that robots bestow upon the human race. They represent our ability to use our minds (the *essence* of being human) in order to conquer physical tasks beyond the capacity of our bodies—our elimination of mundane chores and our freedom to pursue more abstract challenges—our mastery of the world.

So while the creators of science and technology have given us actual robots that improve the upward climb of the human race, the creators of science-fiction drama show us that we are almost God-like in our conquest of the Earth—and of all we survey!

To those movie and television artists who have given us this rare and exhalted view of ourselves, this book is dedicated.

Kerry O'Quinn/1979

INDEX

Rare behind-the-scenes photo from *Metropolis* (1927). The robotrix costume is laced-up in the back.

1970 - 1979

SPECIAL COLOR SECTION

TV ROBOTS

1940—1959

1960s

1970s

INTRODUCTION

Although robots are an ancient concept, the word "robot" was invented in this century. It is derived from the Czechoslovakian word *robota*, meaning "drudgery, servitude, or forced labor." It was coined by the brilliant Czech playwright Karel Capek, in 1920.

Capek (pronounced Chop-ek) used the word to describe the central characters of his classic science fiction play, R.U.R. *(Rossum's Universal Robots)*. In the play, an eccentric old scientist named Rossum has invented an artificial substitute for flesh and bones through biological and electrical means. He uses this substance to pursue his dream: the creation of artificial life-forms. It is his hope that, by so doing, he can make the existence of God an irrelevant question and prove that Man—with the aid of science—is truly the master of his world.

Rossum's experiments end in aborted atrocities and he loses his sanity. His nephew, a "modern-thinking" engineer, resolves to create something practical from his uncle's experiments. He sets about making the perfect manual labor-machine. Using Man as his model, he streamlines the design, eliminating lungs, heart, emotions and soul as unproductive elements.

Young Rossum's experiments succeed beyond his wildest dreams and Rossum's Universal Robots goes into mass production with worldwide distribution to meet an evergrowing demand. Inevitably, the robots are bought in great quantities by warring nations and converted into great, soulless armies. They are totally ruthless and efficient, destroying on command and decimating every living thing in their path—enemy soldier and helpless civilian alike.

Meanwhile, the scientists back at R.U.R. are not unaffected by world events and discuss the possibility of closing down the robot factory. But there are too many counterarguments—closing the factory will put thousands out of work and mean that people will have to go back to doing hard labor and menial jobs; the robots represent humanity's greatest achievements; countries would still war . . . etc. One of the biological scientists, overwhelmed by his guilt, continues to experiment with the formula for the robotic mixture, hoping to effect change.

Ultimately, the robot armies decide on rebellion. They perceive humans as flawed, wasteful and unfit masters. The end of humanity is near; there shall be a purge and the robots shall rule the Earth. But they cannot procreate: reproductive systems do not add to a worker's efficiency and so are not part of the robots' design. The robots must storm the factory and take over production to insure their continuance.

The last remaining humans, those at R.U.R., argue over the secret formula. The robots have said that they will allow them to live if the formula is turned over. Some are willing, others would die first. At this point, there emerges on the stage two new domestic robots, still functioning faithfully within the factory. They are Primus and Helena and they are deeply concerned with the fate of the other, for they are in love. They are, in fact, the product of the additional fiddling that was done to the formula. That formula, it becomes clear, has been accidentally destroyed, and so the rebellious robots are destined to die out within a generation. But their work is done—they have taken the whole of humanity with them. In its place shall come a new kind of creature to inherit the Earth; the offspring of Helena and Primus. The play ends with a reading from Genesis, as the new Adam and Eve exit holding hands . . . garlands in their hair.

* * *

The ideas presented and examined by Capek are profoundly disturbing. Yet they accurately reflected the state of affairs of the world of his day. He had just lived through WWI and seen the horrors it had wrought. He was aghast

N.Y. stage production of *R.U.R.*, from the early 1920s.

that technological progress had been perverted to making man a more efficient killer. He witnessed the cold, impersonal horror of mass death and destruction. At the same time, he saw the people of the ghetto—sick and starving, stripped of their dignity and most of their humanity, willing to kill for a crust of bread.

All of this and more went into the shaping of R.U.R. For the vehicle to properly express his concern, for the embodiment of the conflicting forces at play, Capek found historical precendent.

Prague, the city of Capek's birth, has a long and rich history of legend and folklore. Capek found his inspiration in one of the most famous of the legends, that of the Golem. "Golem" is the Hebrew word for "embryo." In the 16th century, rabbi Judah Loew is said to have created a man out of clay to protect the Jews of Prague from their enemies. This synthetic monster, larger by far than a man, was brought to life by placing a paper inscribed with the mystic "shem" (one of the names of God) in its mouth. There are various endings to the legend. One states that ultimately the Golem rebelled and brought grief to the house of his "creator" because he was used for mundane purposes and not the work of God. The Golem was a powerful tool—a gift of God—and needed a wise, unselfish and experienced mind and soul to guide its actions.

This 16th century legend is similar to another story of an artificially created life-form; one that was written in the early 19th century and one with which Capek was also familiar.

Mary Shelley was greatly disturbed by the swift "progess" and technological invention that was visibly altering her world. The impetus of the Industrial Revolution had made men bold. It led them to experiment with forces of nature that were previously considered to be strictly the province of God. It was at this time that Benjamin Franklin was conducting his famous experiments with lightning and the harnessing of electricity. In 1816 Shelley had written a novel that expressed her fears for the soul of mankind; she had called it *Frankenstein*. There are some reports that a second, revised edition quickly followed after the news of Dr. Franklin's experiments had reached her, and it was there that the monster's life-force was first portrayed as generated by a bolt of lightning.

Mary Shelley was (as Capek was to be after her) influenced by the legend of the Golem as well as the world in which she lived. Still, there is another legend that also contributed to the formulation of *Frankenstein*, for the novel is subtitled *The Modern Prometheus*.

The legend of Prometheus is found in classic Greek mythology. Prometheus was a Titan—one of the elder gods of Mt. Olympus. He is credited with creating mankind, forming the first people out of clay and giving them the spark of life. It is also said that though they lived, his creatures faired badly indeed, and were in danger of dying out completely. In an effort to save them, Prometheus stole *fire* and *the arts* from the heavens and gave them as gifts to mankind. Zeus punished him for this deed by having him chained to a rock where a vulture would come to pluck out his liver from his chest and devour it. Zeus would then see to it that Prometheus' liver grew back overnight and the vulture would return the next day to perform the torturous punishment all over again—for eternity.

Opening scene from *R.U.R.*, N.Y. production.

It is interesting to note here that the liver is the organ that manufactures red blood cells; the fluid of life. The Greeks were aware of this and so Prometheus' punishment must be seen as a constant reminder that the gift of life is one that only those gods properly empowered may give. Also of interest is the English translation of *Prometheus*. It means, literally, "before" *(pro)*; "to learn, or to think" (*mathein:* also the root of the word *mathematics*). Hence, forethought. A possible interpretation of this is that the Greeks were saying that Prometheus already knew the dire consequences of his actions before he acted. And yet, in an effort to keep his creation alive, he threw this knowledge and all caution to the wind.

Prometheus had a heavenly counterpart, according to the mythology. It was Hephaestus, the lame god of fire and the arts. But Hephaestus did not use his powers to create creatures who likened themselves to the gods—he created mechanical servants.

Hephaestus is said to have made the bronze giant Talos in his heavenly forge. It was given to King Minos, who set it to guarding the shores of Crete. This the giant did mercilessly and efficiently, until Jason and his Argonauts came upon that land. The gleaming giant had one fatal vulnerability: Hephaestus had filled its huge metallic body with vital fluids to give it life and had plugged the opening with a stopper. Medea managed to charm the stopper out and the precious liquid ran out of Talos—taking the spark of life with it.

The mythology surrounding Hephaestus also claims that he created several golden, mechanical women to serve his quests and assist him in his experiments.

These then, are the historical precedents, both fact and fancy, upon which is built the continuing mythology of the robot: Prometheus, Hephaestus, the Golem, the Industrial Revolution, Frankenstein, R.U.R.

The basic tenets of that ongoing mythology can be formally set down as follows:

1. He who creates life artificially acts outside the province of man.
2. An artificial "living" creature is, being man-made, a soulless creation.
3. It is impossible to create a "perfect" artificial being because humans themselves are flawed. Even if such a creation gives the appearance of perfection, defects will soon appear.
4. An artificial creation is ultimately not responsible for its actions. That responsibility falls upon its creator(s).
5. The gift of life is not man's to bestow. The attempt will always lead to the downfall of the giver.
6. Once an artificial being becomes self-aware, it has the chance of acquiring a soul.
7. Once an artificial being becomes self-aware, it is not to be treated as an inferior or used as a beast of burden.
8. Ultimately, all attempts at producing a perfect form of artificial life will end in death and destruction.
9. The mistakes of Prometheus, Rabbi Loew, Dr. Frankenstein and young Rossum, if they are not learned, are destined to be repeated.

* * *

Even before the word "robot" had been invented, it had been the subject of several early, silent films. Ten films produced before 1920 had a robot or robots as their theme. It was alternately presented as a mechanical, mindless, labor-saving device *(Motor Valet, Electric Servant, the Mechanical Man)* and something a bit more personally satisfying *(Mechanical Husband, Electric Doll, Mechanical Mary Anne)*. In almost each instance, whether presented as a tin man, an automated box or an accurate reflection of flesh and blood, somehow it malfunctioned and either *destroyed* itself and/or its owner/creator.

In the decades of the thirties and forties, Hollywood turned the tin can robot into a standard "B" melodrama prop. Usually in the control or employ of some sinister character, the robot was used for plunder, kidnap, murder and/or to menace innocent young girls. Sometimes it was the invention of the good guys—who only used it as a super-servant of mankind—but stolen by the bad guys. Whoever controlled it could command the mindless mechanical man to do its bidding. This was generally accomplished by giving voice orders; either over a microphone to control it at a distance, or by speaking aloud in the robot's presence. Then again, there were several films from this period that merely used robots as background scenery, totally superfluous to the plot.

The thirties and the forties also saw numerous remakes of the early robot-theme movies—*Frankenstein, The Golem, The Mechanical Handy Man, The Tin Man*—and their inevitable portrayal of machines gone haywire, causing grief and sometimes death to come to their creators.

These continued to be the main themes surrounding robots in film until 1951, when *The Day the Earth Stood Still* premiered. Gort, the gleaming, seamless, indestructible metal giant is shown to be mentally superior to humans, capable of sophisticated compassion for an alien race, and totally dedicated to its humanoid companion, Klaatu. This robot from the outer reaches of space was capable of dealing out considerable destruction and probably would have, if it had followed the typical Earth (read that *Hollywood*) mold.

From *The Golem*, 1914 silent film.

Two important events took place in the early '50s that were to eventually change that archetypical concept of robots. One occurred in the realm of science. It was the development of the budding computer industry. Already machines had been made to serve man in his pursuit of knowledge and truth—and economic advantage. Computers were establishing the reputation of being efficient, correct, emotionless and tireless tools. It was a great breakthrough. These machines were benign; they did only—*and exactly*—what they were programmed to do.

The second event occurred within the literature of science fiction. In 1950, a collection of robot stories written by Isaac Asimov were published in one volume under the title of *I, Robot*. It contained the now canonized "Three Laws of Robotics":

1. A robot may not injure a human being, or, through inaction, allow a human being to come to harm.

2. A robot must obey the orders given it by human beings except where such orders would conflict with the First Law.

3. A robot must protect its own existence as long as such protecton does not conflict with the First or Second Law.

These laws were built into the components of any robot with a "Positronic brain." This was another creation of Asimov's—an incredibly small, super-sophisticated brain package that freed robots from a physical or mental dependence on computer information storage banks. In a word, it made them *mobile*. And, because they were governed by the Three Laws, Asimov made his robots self-aware and cognizant of their positions in the world—that is, what they were designed to do and what was expected of them.

These developments gave rise naturally to the next step in movie robots: Robby. In *Forbidden Planet* (1955), one of the central characters is a robot that, being an artifact of a highly sophisticated alien science, is obviously superior to humans. Despite his potential destructiveness he presents no physical threat, for he is governed by a set of programs that correspond quite closely to Asimov's Three Laws.

Ironically, though, it was the first development—the perfecting of computer technology—that would have the more pervasive influence on cinematic robots. When robots were only a theoretical possibility, filmmakers exercised liberal poetic license and produced a fantastic assortment of ambulatory tin cans and, when budgets ran low, androids. (An android is a robot built to look like a human.)

But when reality caught up to theory, a different picture emerged. Scientists were perfecting the one robotic component that the movies for the most part neglected—its brain. Not only that, they seemed to have no inclination to perfect human-like mobile mechanical bodies for these super-brains, leaving them instead in giant metal boxes. Those robotic machines capable of movement that were invented were also very stupid—that is, they could only perform one or two functions.

In 1954, the current scientific trends were combined with a Cold War scenario and the classic robotic pitfalls to produce *GOG*—also known as *GOG The Killer*. Here a giant, immobile computer spawns two ambulatory assistants, Gog and Magog. The three machines are working on the U.S.'s first space station project from a hidden, desert base. Enemy agents sabotage Novac, the computer, causing Gog and Magog to go berserk and destroy the scientific station . (The two units are distinctly machine-like and are quite possibly the progenitors of Dr. Who's most dangerous enemies, the Daleks.)

It took the combined genius of Arthur C. Clarke and Stanley Kubrick to project the development of computer technology and robotic mobility into the 21st century, in *2001: A Space Odyssey*. They created the ultimate thinking, feeling, exploring, computing machine, HAL 9000, and the perfect physical housing to take it into the outer reaches of unknown space, the starship *Discovery 1*. HAL's is the brain that controls the starship. And it is no accident that the craft resembles some gigantic, metallic brain and backbone—with its spherical control center and vertebral-shaped cargo containers strung out in back of it. When HAL's lethal plot to eliminate his human passengers goes haywire, he tries to reason with Dave (Kier Dullea) by logically pointing out that *he* (HAL) is perfectly suited to the mission at hand and that the presence of humans represents a liability. But it turns out that HAL has been but a sophisticated tool of humanity's after all, in helping the species achieve its destiny among the stars. *2001* still stands as the most powerful public prediction that humanity's development is now irretrievably linked to its technological progress and the ability to co-exist with it.

* * *

Science in the seventies has brought about another surprising turn of events in developing robotics. The terms "cyborg" and "cybernetics" have become part of mainstream vocabulary. Webster's New World Dictionary defines a cyborg as: "A human being modified for life in in a nonearth environment by the substitution of artificial organs and other body parts." Cybenetics is de-

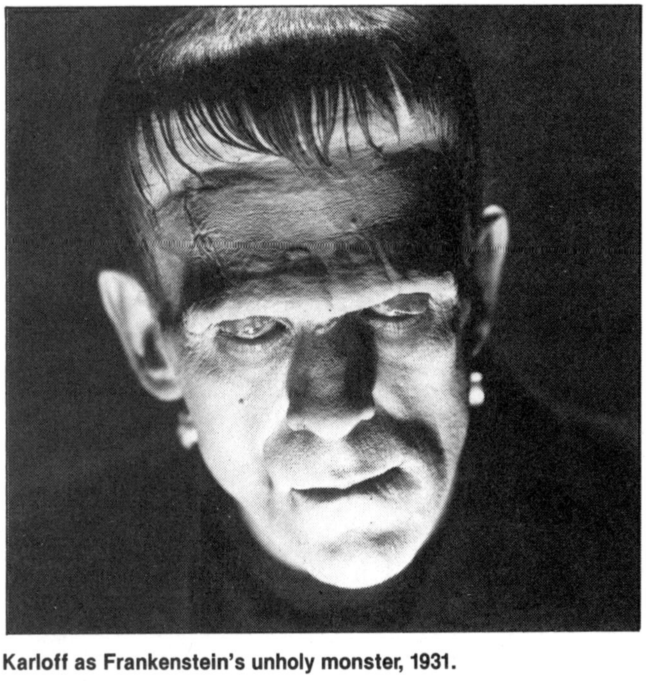

Karloff as Frankenstein's unholy monster, 1931.

fined as: "A science dealing with the comparitive study of complex computers and the human nervous system." But to the society-at-large, "cybernetics" meant the arcane practice of making computers human, while "cyborg" meant Steve Austin—better known as TV's *Six Million Dollar Man*—a NASA astronaut who has undergone a traumatic accident and has had his body re-made with several superior, artificial parts.

The subtleties achievable by computer brains and the scientific investigation to find an exact definition for the term "life" were the inspiration for the next evolutionary step taken by cinematic robots, although there is a certain amount of *deja vu* in their conceptions.

In 1970, *Colossus, The Forbin Project* was released. Six years later, *Demon Seed* hit the silver screen. These two films further refined the legacy of *Frankenstein* and *R.U.R.* In *The Forbin Project*, an American super-computer, Colossus(built for military defense) hooks up with its secret Soviet counterpart, Guardian. The two brains confer and decide that the human race is in a state of chaos—on the brink of self-inflicted oblivion. Disregarding the cost in terms of human life, the machines eliminate or neutralize their heaviest opposition and proceed to dictate living terms to humanity. Having perceived the pitfalls of human flaws, they have decided that super-computers are far better suited to oversee the day-to-day life and future development of the human race. At the end of the film, Colossus speaks to Forbin, its creator: "In time you will grow to love me . . . to worship me. . . ."

In *The Demon Seed*, real research in such varied scientific disciplines as recombinant DNA, computer technology and cybernetics are merged into a mesmerizing, horrifying scenario.

As a super-computer becomes cognizant of itself and its various functions and interconnections, it takes over the computer terminal and mechanized servants and research tools that its creator has built into his house. After doing so it realizes that it has a captive subject (Julie Christie, the doctor's wife), and an experiment to perform that has never been programmed by any human agent. The computer wishes *to mate* with its captive human female and produce a *living*, ambulatory offspring.

In short, it succeeds, bringing forth from an enhanced incubator a flesh-and-blood reflection of itself. The movie ends with the hybrid opening its too-human eyes, surveying the situation and pronouncing (in the same chilling, emotionless tone as the computer) two terrifying words: "I live."

* * *

Cinematic robotics have also been developing along the lines of Asimov's creations, despite what's going on in "the real world." *Star Wars* featured a pair of "droids" that are the creations of a technology obviously governed by some version of the Three Laws of Robotics. C3P0 and R2D2 are totally devoted to their human masters and apparently have no choice when instructed to carry out difficult tasks, even when their own safety is endangered.

Paul Wegener as *The Golem*, 1914.

However, the development of celluloid robots continues to be influenced by the reality of the day. NASA scientists have perfected yet another kind of robotic servant—the planetary and interplanetary probes. The major adversary that the crew of the new *Enterprise* must defeat (in the soon-to-be-released *Star Trek* movie) is a ghastly, mutated version of a 20th century interplanetary probe from NASA's Voyager series.

What will the next step be? Who can say. . . . The only pertinent piece of advice is to read about the latest scientific breakthroughs—and think about how you might apply them to future robotics.

In the meantime, film robots—whether they are super-computers, ambulatory tin cans or human-like androids—will continue to follow the classic scenarios of Prometheus, *Frankenstein* and *R.U.R.*: would-be creators of artificial life take note, and beware.

H.Z.

* * *

Note: Several films for which there are no surviving prints have been included in the listings because of their historical importance. Also, many films that are merely dreary variations on a well-played theme have been excluded. For example, only three *Frankenstein* films have been listed: the original, 1910 version filmed by Thomas Edison, the 1915 silent, *Life Without Soul* and the classic 1931 treatment starring Boris Karloff as the monster.

ROBOTS ON FILM

1897-1919

THE CLOWN AND THE AUTOMATON
(France), 1897, B&W, 65 feet, silent.
The first known motion picture involving a robot, made by the legendary film pioneer, George Melies.

COPPELIA, THE ANIMATED DOLL
(France), 1900, B&W, 130 feet, silent.
Another early film directed by George Melies, based on the ballet "Coppelia" by Delibes. The story is that of a robot-like doll who becomes a woman.

THE MOTOR VALET
Alpha Trading Co. (Great Britain), 1906, B&W, 200 feet, silent.
A rather volatile early film short: a robot-servant goes berserk, smashes up the furniture and then explodes.
Director: Arthur Cooper.

THE DOLL'S REVENGE
Hepworth (Great Britain), 1907, B&W, 225 feet, silent.
A young boy (Bertie Potter) breaks a doll belonging to his sister (Gertie Potter). It mends, grows, tears him up and eats him.
Director: Lewin Fitzhamon.

AN ANIMATED DOLL
Essanay, 1908, B&W, 750 feet, silent.
A little boy steals the robot doll of a neighborhood inventor and gives it to his sister for a birthday present. The doll dances for them until the inventor arrives and retrieves it. All their entreaties to the inventor to let them keep it are in vain, and the little girl cries herself to sleep, clutching her own small doll.

THE AUTOMATIC SERVANT
Urban-Eclipse, 1908, B&W, 367 feet, silent.
An inventor has developed a life-size robot figure, and has invited all his friends over to view the marvel. But a servant breaks the mechanism while the inventor is out, and he is forced to substitute his human valet for the robot to save face. The valet does his best to simulate mechanical movements, and even pours wine down his master's neck instead of into his glass at dinner for verisimilitude.

THE ELECTRIC SERVANT
Urban Trading Co. (Great Britain), 1909, B&W, 440 feet, silent.
Another robot gone mad and breaking up furniture. This time the robot belongs to Professor Puddenhead, and as usual it goes berserk and wreaks havoc.
Director: W.R. Booth.

THE AUTOMATIC MONKEY
Gaumont (Great Britain), 1909, B&W, 324 feet, silent.

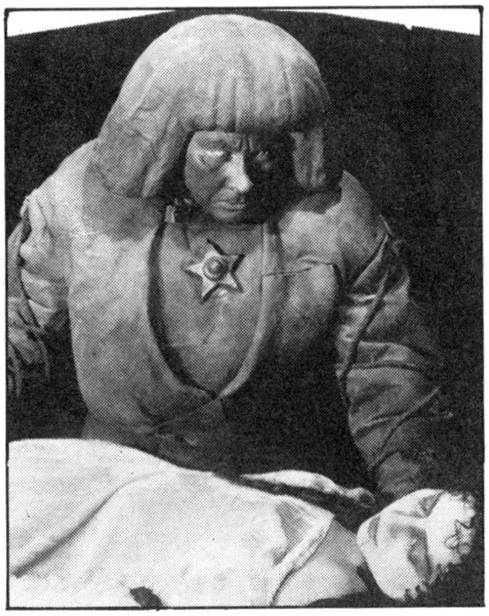

From the 1920 remake of *The Golem*.

A French film, released by England's Gaumont. Jocko, a full-size robot monkey, paints and sculpts, does a little architecture and proves to be a student of music. Harmless fun—if you like monkeys.

FRANKENSTEIN
Edison, 1910, B&W, running time unknown, silent.
A lost film, based on the novel by Mary Shelley, starring Charles Ogle as the monster.
Director: J. Searle Dawley.

A MECHANICAL HUSBAND
London Cinematograph (Great Britain), 1910, B&W, 415 feet, silent.
A girl objects to her father's choice of a husband for her and falls in love with an automaton.
Director: S. Wormwald.

THE MECHANICAL MARY ANNE
Hepworth (Great Britain), 1910, B&W, 560 feet, silent.
A clockwork maid wrecks the house.
Director: Lewis Fitzhamon.

THE AUTOMATIC MOTORIST
Kineto (Great Britain), 1911, B&W, 610 feet, silent.
A robot-chauffeur drives two honeymooners under the sea and then to Saturn.
Director: W.R. Booth.

THE DOLL'S REVENGE
Clarendon (Great Britain), 1911, B&W, 410 feet, silent.
A young girl's robot doll plays tricks on the girl's brother and finally trips him into the bath water.
Director: Percy Stow.

HOW THEY WORK IN CINEMA
Eclair (France), 1911, B&W, length unknown, silent.
Reflecting a wish of directors through the ages: a film director uses robots to make movies instead of human actors. Strictly tongue-in-cheek.

AUTOMATON OR ACROBAT?
Pathe (France), 1911, B&W, 414 feet, silent.
Supposedly a robot, the film actually features an acrobat simulating a mechanical toy.

COPPELIA
(Denmark), 1912, B & W, length unknown, silent.
Another early version of the story of the dancing robot doll; a popular subject since it requires no costly special effects.

THE ELECTRIC DOLL
Cricks (Great Britain), 1914, B&W, 485 feet, silent.
An inventor's robot causes trouble when

The mad Rotwang animates a robotic Maria in *Metropolis*, 1927.

it is mistaken for a corpse.
Director: Edwin J. Collins.

THE GOLEM (aka THE MONSTER OF FATE)

Bioscop (Germany), 1914, B&W, length unknown, silent.

In this first film version of the Golem story, Paul Wegener stars as Rabbi Loew's man of clay. Created to protect the Jews in times of trouble, the Golem wanders off and goes on a soulless rampage before being rendered inanimate again.
Directors: Paul Wegener, Henrik Galeen. Screenplay: Henrik Galeen, based on the novel by Gustav Meyrink. Art Director: Robert A. Dietrich. Supporting Players: Henrik Galeen, Albert Steinruck, Ernst Deutsch, Lydia Salmanova.

TALES OF HOFFMAN

Union (Germany), 1914, B&W, length unknown, silent.

An early version of the story of a man who falls in love with a robot dancing doll.
Director: Richard Oswald. Screenplay: Fritz Friedman and Richard Oswald, based on the stories of E.T.A. Hoffman. Art Director: Manfred Noa.
Cast: Warner Krauss, Erich Kaiser-Tietz, Kathe Oswald, Paula Ronay.

THE DANCING DOLL

Kalem, 1915, B&W, 3 reels, silent.

The story of a robot doll that can walk and dance. And dance. And dance. And dance. And dance. . . .
Screenplay: Howard Irving Young.
Cast: Vivian Wessell, Wayne Nunn, E.T. Roseman, George Moss, Harland B. Moore.

LIFE WITHOUT SOUL

Ocean Film Corp., 1915, B&W, 5 reels, silent.

The second film based on the Frankenstein legend. Dr. Frankenstein creates a monster (Percy Darrell Standing) and it pursues him all over Europe.
Director: Joseph W. Smiley.
Screenplay: Jesse J. Goldburg, based on the novel by Mary Shelley.
Supporting Players: William A. Cohill, Jack Hopkins, Lucy Cotton, George DeCarlton, Pauline Curley, David McCauley, Violet DeBiccari.

THE MECHANICAL MAN

Joker, 1915, B&W, 1 reel, silent.

An inventor creates a robot man with controls on its back; when it breaks down, another man is forced to impersonate it.
Director: Allen Curtis.
Screenplay: Clarence Badger.
Cast: Max Asher, Phroso.

THE GOLEM

(Denmark), 1916, B&W, length unknown, silent.

Danish version of the Czechoslovakian legend of a clay figure bought to life through magic and used unwisely.
Director: Urban Gad.

HOMUNCULUS (aka HOMUNCULUS THE LEADER)

Bioscop (Germany), 1916, B&W, serial in 6 parts, silent.

Homunculus, a soulless, artificial man of great intellect, becomes dictator of a large country. He plans to conquer the world, but is destroyed by lightning. A variation on Shelley's theme.
Director: Otto Rippert. Screenplay: Otto Rippert and Robert Neuss, based on the novel by Robert Reinert. Art Director: Robert A. Dietrich.

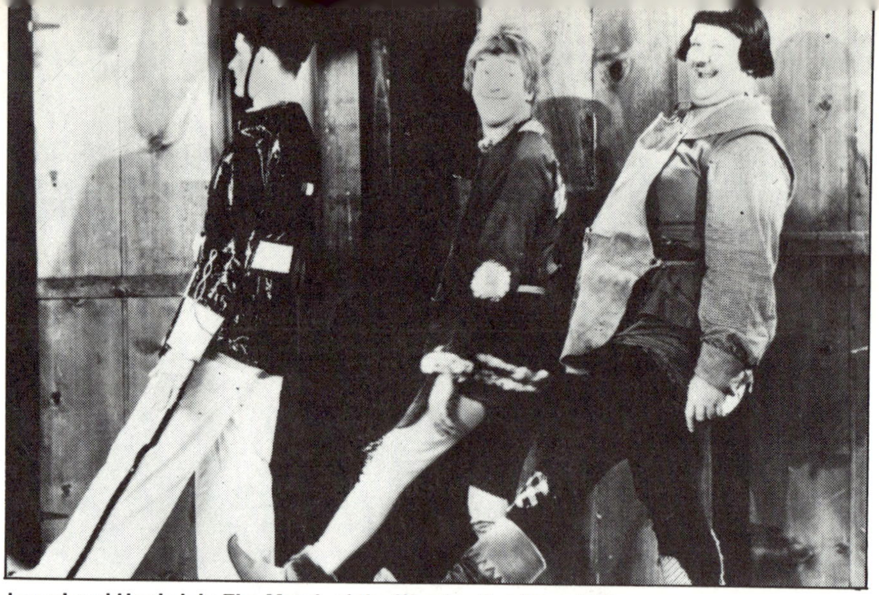

Laurel and Hardy join *The March of the Wooden Soldiers*, 1934.

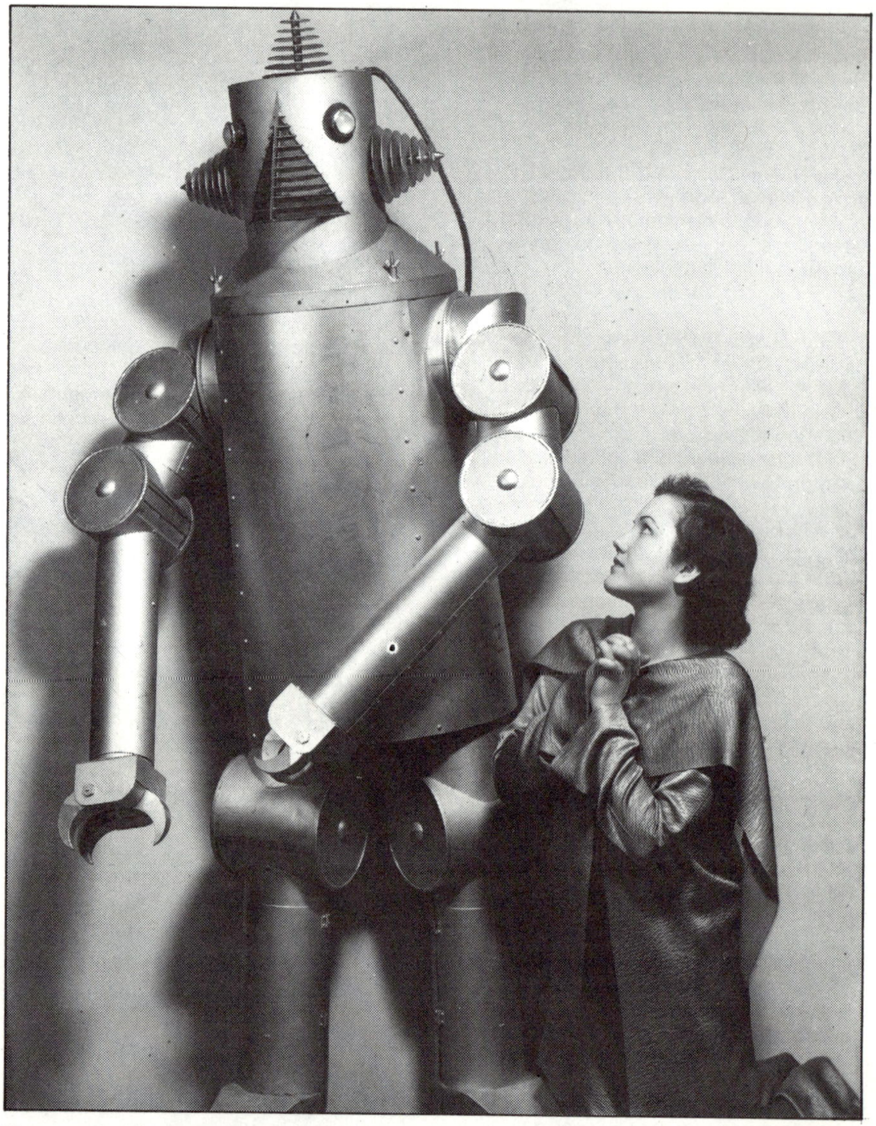

Tin can robot from *The Vanishing Shadow*, 1934.

THE GOLEM AND THE DANCING GIRL

Bioscop-Union (Germany), 1917, B&W, length unknown, silent.

Actor Paul Wegener, appearing as himself, puts on the Golem disguise to fool and impress a dancer.
Director: Paul Wegener. Supporting Players: Lydia Salmanova, Rochus Gliese.

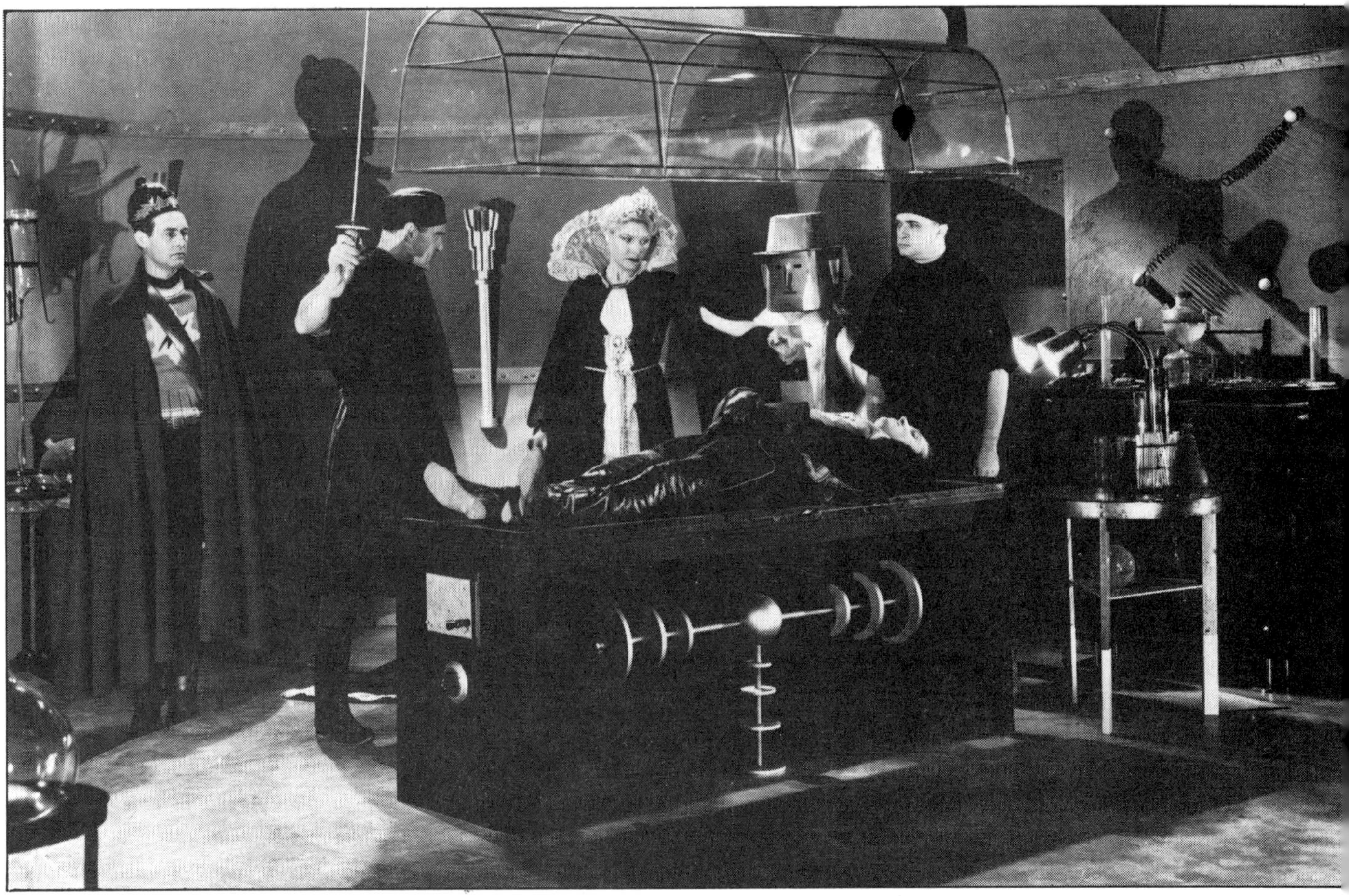

Gene Autry's in trouble in *The Phantom Empire,* 1935.

ALRAUNE

Phoenix (Hungary), 1918, B&W, running time unknown, silent.

This is the first film adaptation of the novel of the same name, by Hanns Heinz Ewers; ultimately, it would be filmed five times over a period of 34 years. Another version was supposedly also made in 1918 by Luna-Film in Germany, but all that survives of it is the title. Even this version seems to be one of those early lost films, with no prints in existence. The book is the story of a scientist who creates a soulless being through artificial insemination, using a hanged criminal and a prostitute. In this first film version the story was changed so that the girl was fathered by a mandrake root. The child grows to adulthood without feelings, learns of her unnatural roots and meets a tragic end.

Director: Mihaly Kertesz (Michael Curtiz) and Odor Fritz. Screenplay: Richard Falk. Cast: Guyla Gal, Rozsi Szollosi, Jeno Torzs, Margit Lux, Kalman Kormendy, Geza Erdelyi, Andor Kardos, Violette Szlatenyi, Karoly Arnyai, Boske Malatinszky.

ALRAUNE AND THE GOLEM

Riesenbioskopfilm (Germany), 1919, B&W, length unknown, silent.

An early example of combining the stories of two famous monsters in one film, some years before Universal began letting Abbott and Costello meet everybody under contract.

Artistic Supervisor: Nils Chrisander. Cinematographer: Guido Seeber.

Bella Lugosi has a trick up his sleeve, as *The Phantom Creeps,* 1939.

Creation of Olympia; *Tales of Hoffman,* 1951.

The Perfect Woman (1949) turns out to be flesh and blood.

THE DOLL
PAGU (Germany), 1919, B&W, length unknown, silent.
A very simple film: a doll is brought to life. But how does the doll really feel about that? There are no prints, no record.
Director: Ernst Lubitsch. Screenplay: Hanns Kraly, based on a story by A.E. Willner, Art Director: Kurt Richter
Cast: Ossi Oswalda, Gerhard Ritterband, Max Kronert, Herrmann Thimig, Victor Janson.

THE MASTER MYSTERY
Octogon Films, 1919, B&W, serial in 15 parts, silent.
Harry Houdini (himself) runs up against a robot with a human brain (Floyd Buckley), but finally reveals it as a fake.
Director: Burton King.

1920-1929

THE GOLEM
Riesenbioskopfilm (Germany), 1920, B&W, 6 reels, silent.
A re-make of the 1914 German film, again starring Paul Wegener as the Golem, and featuring several other members of the original cast.
Directors: Paul Wegener, Carl Boese. Supervising Director: Enrst Lubitsch. Screenplay: Paul Wegener, Henrik Galeen. Supporting Players: Albert Steinrack, Ernst Deutsch, Lothar Mathel, Lydia Salmanova, Otto Gebahr, Greta Schroeder.

LA POUPEE
Milton (Great Britain), 1920, B&W, 5200 feet, silent.
A young man marries a girl, believing that she is a robot-doll; presumably he is pleasantly surprised on his wedding night to discover that he doesn't have to wind her.
Writer-Director: Meyrick Milton. Screenplay based on the opera by Edmon Audran, from the story "The Sandman" by E.T.A. Hoffman. Cast: Flora Le Breton, Fred Wright, Richard Scott, William Farren, Gladys Vicat.

THE GOLEM'S LAST ADVENTURE
Sascha Films (Austria), 1921, B&W, length unknown, silent.
As later films would show, this was actually only the beginning. An early lost film.
Director: Julius Szomogyi.

THE MIRACLE OF TOMORROW
Piel Films (Germany), 1923, B&W, length unknown, silent.
Henry Piel not only released this film, he also directed and starred in it; it's the story of an electrically-controlled robot.

AELITA: THE REVOLT OF THE ROBOTS
Mezharabpom (Soviet Union), 1924, B&W, 45 min., silent.
Often called "The Russian R.U.R.", this film carries the Russian Revolution to Mars through two cosmonauts, Los and Busev. They land on Mars and discover the Martian people to be scantily clad types enslaved by having their heads encased in what look like cardboard boxes with large oval-shaped metal ears attached. They are ruled over by Queen Aelita (Julie Solnzewa). Los (Nikolai Tserectelli) falls in love with the Martian queen, but Busev (Igor Yinski), a more political type, sets out to free the downtrodden masses from their boxes, and starts a revolution. A number of robots appear, having square blocklike hands, articulated joints much like medieval armor, and heads which greatly resemble Pickering stereo cartridges with construction helmets thrust through them. A Martian woman appears at one point wearing what looks like the mechanism from a folding umbrella.
Director: Yakov Protazanov. Screenplay: Fyodor Otzep and Alexei Falko. Production Designer: Isaac Rabinovitch. Costumer: Alexandra Exter of the Kamerny Theatre of Moscow. American Titler and Editor: Benjamin de Casseres.

A MACHINE THAT THINKS
Bray Films, 1924, B&W, 1 reel, silent.
Ah, but what did it think about? We do not know.
Producer-Director: John Bray.

METROPOLIS
UFA (Germany), 1927, B&W, 17 reels, silent with musical score.
By the year 2000, the huge skyscraper city of Metropolis created by Master Industrialist Jon Frederson (Alfred Abel) is divided into two distinct classes: the workers live underground in slums and labor unceasingly to run the machines that power the city, while the elite live aboveground in sun-drenched luxury. A revolt of the workers is brewing, led by Maria (Brigitte Helm), an idealistic young woman who has become involved with Frederson's hitherto socially unconscious son, Freder (Gustav Frohlich). Frederson decides to teach the rebellious workers a lesson about the folly of resisting the established order: he kidnaps Maria and orders the mad scientist Rotwang (Rudolph Klein-Rogge) to build a robot duplicate of her. The robot is a metallic, mechanical woman until Rotwang transforms it into a mirror image using Frankensteinian equipment. The robot is sent below to incite the workers to destroy the dikes, an act which also destroys many of their homes and kills many of their children. The riot that Frederson wanted to incite quickly gets out of hand and spreads to the surface: Rotwang is killed by Freder, Maria is freed, and much of the city is destroyed. Frederson is forced to enter into a partnership with the workers to rebuild the city for the benefit of all the people, not just the privileged few. The film is in many ways a reflection of the conditions

that existed in Germany in the mid-20s; sadly, Germany's Maria turned out to be Adolph Hitler.
Director: Fritz Lang. Screenplay: Fritz Lang and Thea von Harbou, based on von Harbou's novel. Art Directors: Otto Hunte, Rich Kettelhut, Karl Vollbrecht. Special Effects: Eugen Shuftan. Music: Gottfried Happertz, Konrad Elfers. Supporting Players: Fritz Rasp, Theodor Loos, Erwin Biswanger, Olaf Storm, Hanns Leo Reich, Heinrich Gotho.

THE CHESS PLAYER

Unusual Photoplays (France), 1927, B&W, 82 min., silent.

The villain uses a chess-playing robot to work his evil plans, but the robot checkmates and kills him.
Director: Raymond Bernard. Screenplay: Henry Dupuy-Mazuel. Cast: Edithe Jehanne, Pierre Blanchar, Charles Dullin.

ALRAUNE

Ama Film (Germany), 1928, B&W, 122 min., silent.

The third film version of the novel by Hanns Heinz Ewers, and often thought to be the best. Brigitte Helm played Alraune, created by a doctor (Paul Wegener) through artificial insemination and combining the worst features of both her parents, a hanged criminal and a prostitute. She grows up without feelings, and every man she comes into contact with dies. Finally she learns of her origin and does away with herself. The film is long and slow, and the act of her creation is handled with a delicacy that renders it nearly incomprehensible.
Writer-Director: Henrik Galeen. Music: Willy Schmidt-Gentner. Supporting Players: Ivan Petrovich, Mia Pankau, Georg John, Valeska Gert, Wolfgang Lilzer, Louis Ralph, Hans Trautner.

THE MARCH OF THE MACHINES

(France), 1928, B&W, length unknown (short subject), silent.

Why were they marching? Where were they going? Did they arrive? What did they do when they got there? Why are there no prints of this film?
Director: Eugene Deslaw.
Cinematography: Boris Kaufman.

1930-1939

ALRAUNE (aka DAUGHTER OF EVIL)

UFA (Germany), 1930, B&W, 103 min.

The first sound film version of the Alraune story, with Brigitte Helm repeating her portrayal of the role from the 1928 version and Albert Bassermann playing her creator. Aside from the "improvement" of sound, the film is thought to be much inferior to the 1928 silent version made by Ama Film.
Director: Richard Oswald. Screenplay: Charles Roellinghoff, Richard Weisbach. Music: Bronislaw Kaper. Supporting Players: Agnes Straub, Kathe Haack, Martin Kosleck, Bernhard Goetzke, E.A. Lichs, Henry Bender.

A completed Olympia dances and dances . . . and dances.

The indefatiguable Republic robot appears *almost* menacing in *The Mysterious Dr. Satan*, 1940.

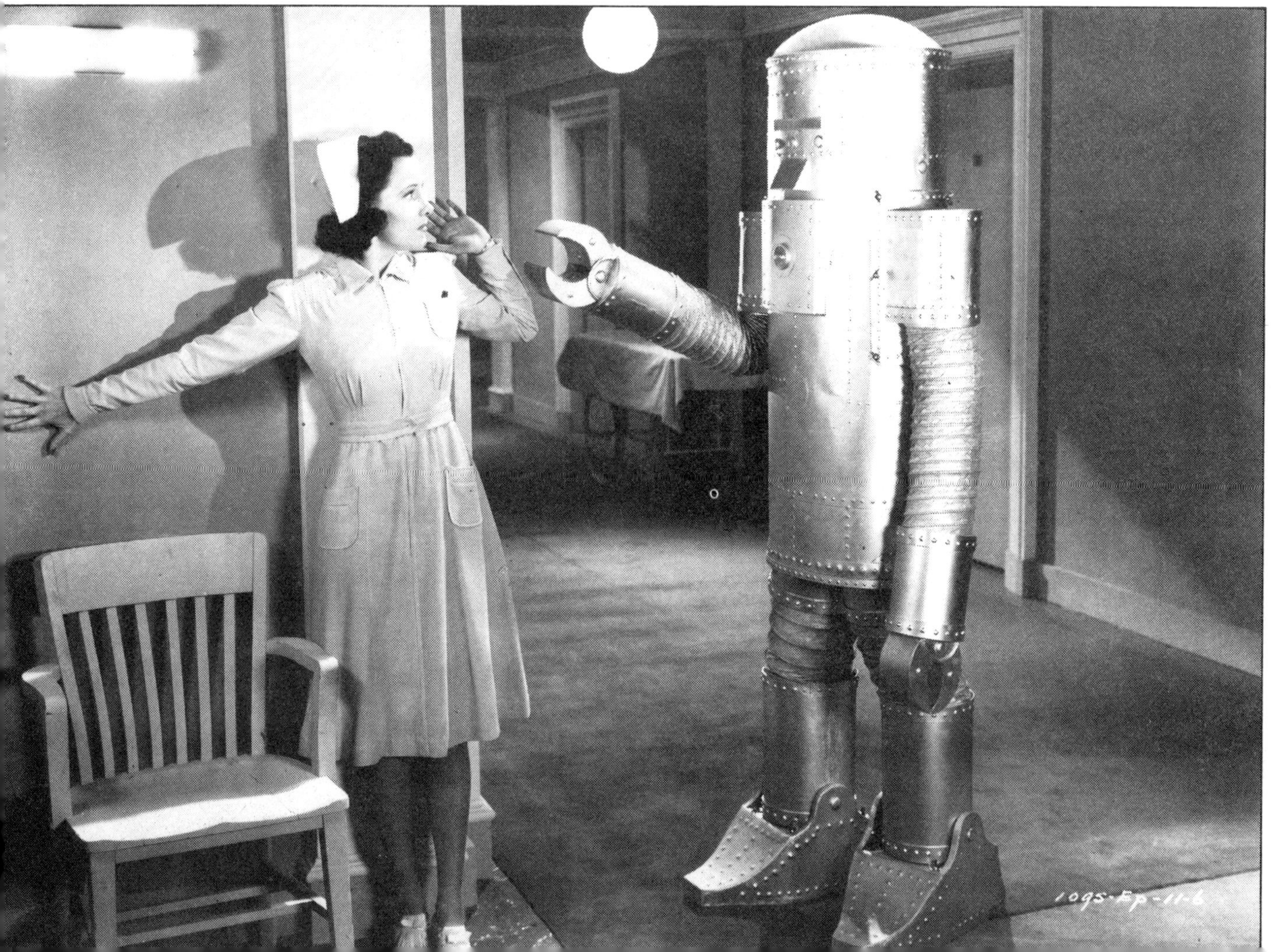

Left and above: from the 1951 *Captain Video* serial, a robot that dies with its hat on (look familiar?).

FRANKENSTEIN
Universal, 1931, B&W, 71 min.
The familiar, classic film which has spawned innumerable, mostly inferior successors and spin-offs. Dr. Victor Frankenstein (Colin Clive) finds success and horror co-mingled in his experiments to create artificial life using electricity. Pieced together from various odds and ends gathered by Dr. Frankenstein on his grave-robbing excursions, the Monster (Boris Karloff) goes on a rampage after mistakenly being given the brain of an executed criminal. Finally cornered by a mob of enraged villagers, the Monster is burned to death in a windmill.
Director: James Whale. Screenplay: Garrett Fort, Robert Florey, Francis Edward Faragoh, John L. Balderston, from the novel by Mary Shelley.
Art Director: Charles D. Hall. Makeup: Jack Pierce. Special Effects: John P. Fulton. Special Electrical Effects: Kenneth Strickfaden. Supporting Players: John Boles, Dwight Frye, Edward Van Sloan, Frederic Kerr, Marilyn Harris, Mae Clark, Lionel Belmore.

ROBOT
Paramount, 1932, B&W, 1 reel.
Animated cartoon about a robot, made by the Fleischer brothers, early animation pioneers and creators of "Popeye the Sailor" and "Koko the Clown."
Producer: Max Fleischer.
Director: Dave Fleischer

ROBOTS
(France), 1932, B&W, length unknown but may have been a short.
French movie about French robots doing French things. You know: smoking cigarettes, eating snails, correcting each other's pronunciation.
Director: Eugene Deslaw.

Star Wars

SPECIAL COLOR SECTION

Robots on Film

Silent Running

Buck Rogers

Invisible Boy

GOG

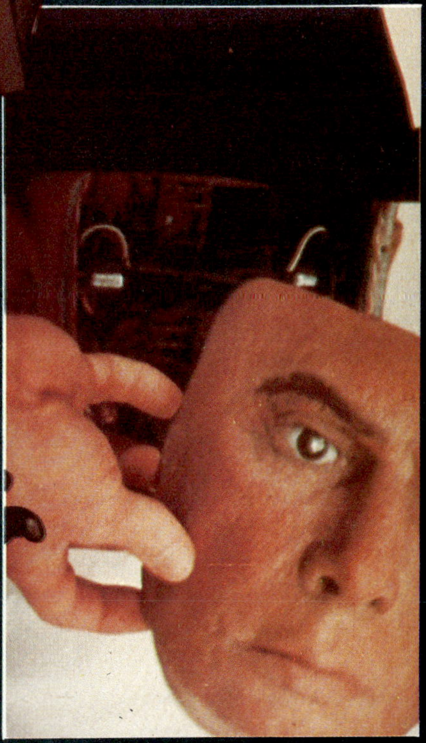

Westworld

Star Wars

The ultimate robot hardware film of all time is *Star Wars.* Robots of all shapes, sizes and functions populate George Lucas' space fantasy landscape; they walk, roll and shuffle around on a variety of legs, wheels and treads. C3PO is a "protocol droid," speaks several alien languages (including English) and has the manners of the perfect British butler. The irrepressible R2-D2 is a "maintainence droid," sophisticated in handling and repairing land-based equipment as well as spacecraft, on the ground and in use.

Above: Production illustrator Ralph McQuarrie's original design for C3PO and R2D2. 3PO is based on the gleaming robotrix from *Metropolis.*

Above: 3PO takes on Chewbacca in 3-D chess game (left) and watches rebel raid develop from the war room (right).
Left: Actor Anthony Daniels takes a break during filming. The C3PO costume is one of the best designed for movie robots. Far left: Luke looks over an R2 unit as D2 patiently waits his turn.

M·G·M PRESENTS
FORBIDDEN PLANET
AMAZING!
STARRING WALTER PIDGEON · ANNE FRANCIS · LESLIE NIELSEN
WITH WARREN STEVENS AND INTRODUCING ROBBY, THE ROBOT
SCREEN PLAY BY CYRIL HUME
PHOTOGRAPHED IN EASTMAN COLOR
DIRECTED BY FRED McLEOD WILCOX · PRODUCED BY NICHOLAS NAYFACK
IN CINEMASCOPE AND COLOR

Everyone's favorite robot, Robby—the most versatile and believable of robots to precede the *Star Wars* era. Above: Robby proves that he cannot kill sentient beings. Below: Robbie's *original* design and his first on-screen appearance.

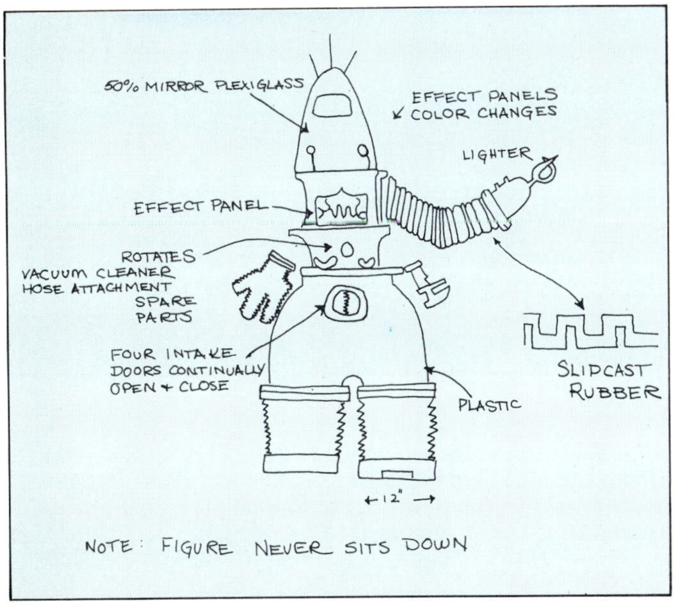

2001: A Space Odyssey

The starship *Discovery 1* is humanity's robot representative to the cosmos.

Astronaut Poole keeps in shape during early part of the voyage. HAL 9000 and the *Discovery* itself take care of all the hard work.

HAL 9000, the ultimate computer, is the brain behind mankind's first attempt to physically contact alien life, by following the Moon monolith's signal to Jupter—and beyond. The *Discovery 1* resembles nothing so much as a giant metallic, human nervous system with the brain attached. Coincidence?

Silent Running

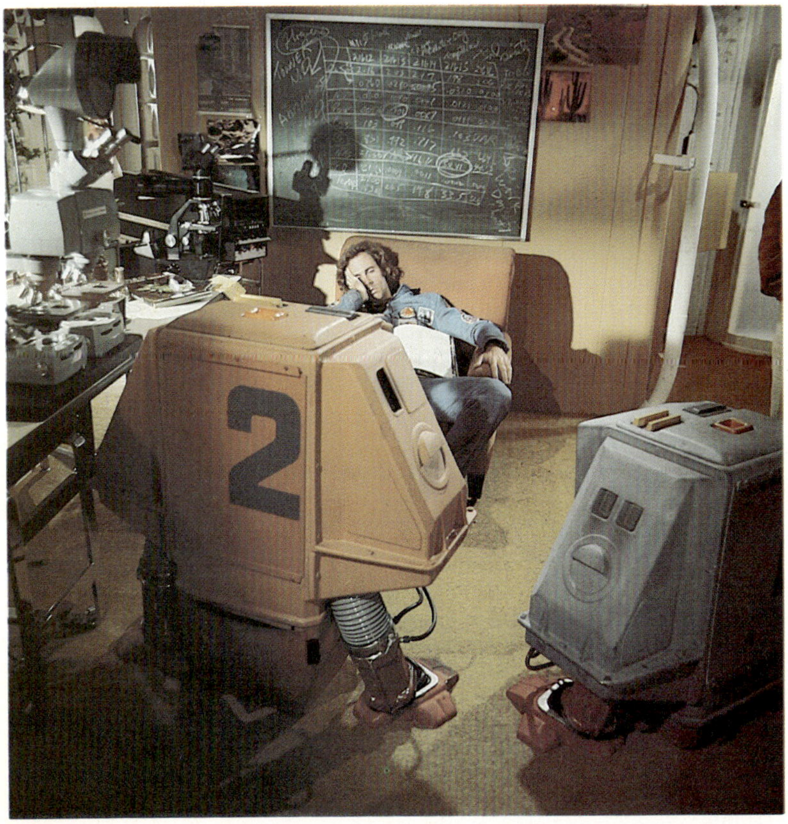

Silent Running is one of the most underrated SF films in recent years. The three robots Huey, Dewey and Louey (Drone 1, Drone 2 and Drone 3) remain among the most believable electro-mechanical servants ever filmed. Multiple-amputee Vietnam veterans were inside of the drones and operated them themselves. These drone units are perhaps the most sophisticated robotic hardware ever developed for cinematic use.

Logan's Run

Roscoe Lee Brown played the mad robot sculptor, Box. At the left, Box works on his sculpture of the nude figures of Logan and Jessica in a love embrace. The scene was cut from the final picture to get the PG rating MGM wanted.

The Day The Earth Stood Still

Gort, the gleaming, seamless metallic robot policeman from *The Day the Earth Stood Still,* is one of the screen's finest robotic conceptions. Armed with world-shattering power, he is capable of enforcing Klaatu's edict to the people of Earth: keep the peace. Still, in the fifties robots had to be pictured as heartless killers to attract an audience—which accounts for the poster art below.

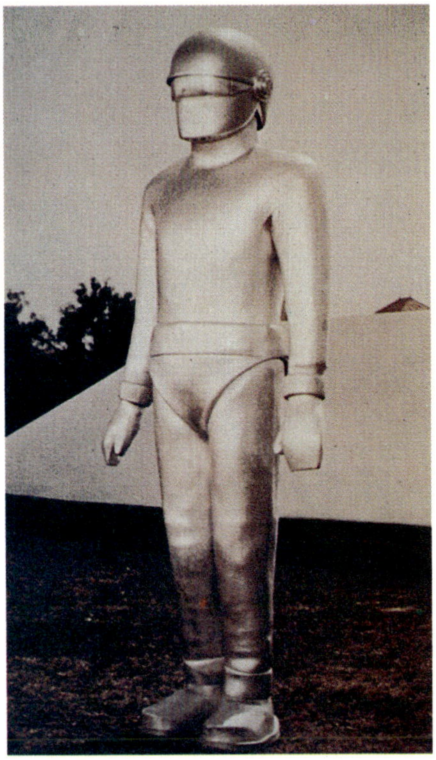

Buck Rogers

No "nirvano gas" this time: the 1979 version of Buck Rogers fell asleep at the wheel—so to speak—of his NASA spacecraft. His robot companion Twiki seems to follow some internal programming of his own and, as C3PO before him, tends to find discretion the better part of valor. Dr. Theopolis (on Twiki's chest) is one of the rulers of 25th century society.

Invasion Earth 2150 A.D.

The Daleks are one of the most popular races of evil aliens ever to be beaten on film. They were so popular, in fact, that after their seeming demise they were brought back in this follow-up to *Dr. Who and the Daleks* and given one more crack at destroying the Earth. But even though they had a force of evil Robo-Men at their disposal this time (in background), Dr. Who remained triumphant.

The Shape of Things To Come

You can't tell the good robots from the bad ones without a scorecard, in this non-sequel to Alexander Korda's *The Shape of Things To Come.* Robot Sparks (below) is wired into evil Emperor Omus' plans for conquest, while the rewired Sparks (right) gets it on for the good guys. He doesn't look quite as imposing as a hero, but he definitely looks more robotic.

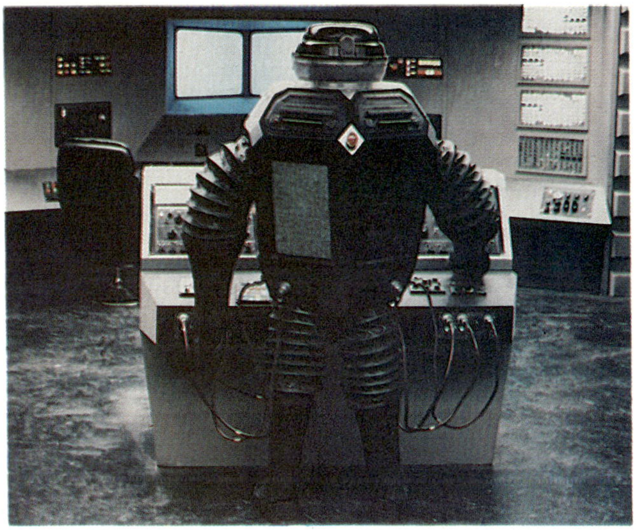

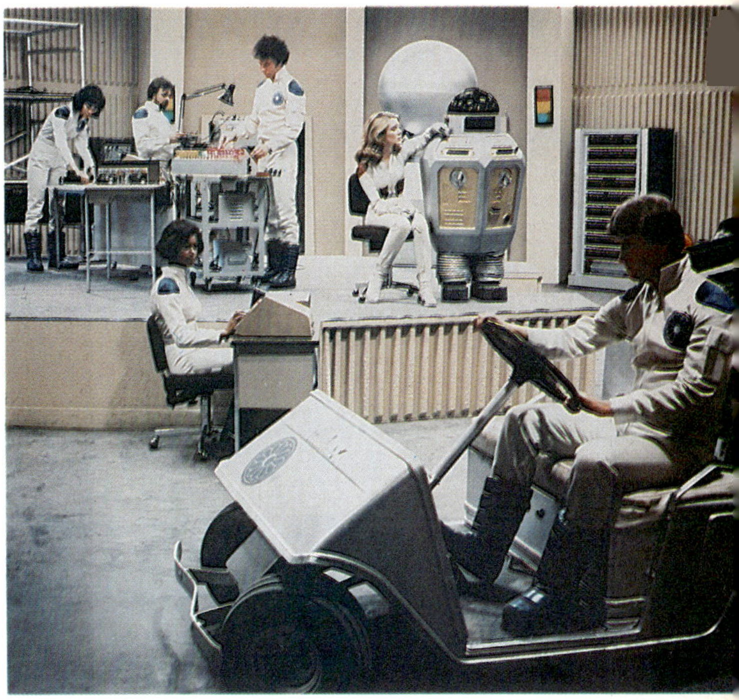

Satan's Satellites

The Republic Robot in all of its awesome glory! This is one of the few times this famous tin can robot ever got the drop on the hero, one-on-one. Most of the time it was used to scare pretty young girls into fainting or screaming.

Westworld

Yul Brynner in *Westworld* played a gunslinger android (ala *The Magnificent Seven*) gone wild. His unstoppable pursuit of the "human" stars is the highlight of the film.

Demon Seed

Proteus is conceived of, and built as, the ultimate thinking computer. As in all films based on this theme, the computer's downfall is inherent in the initial programming given it by its designer. In this instance, it's an insatiable desire for human experience (the better to predict human behavior) that leads Proteus into rape, murder . . . and beyond.

Starcrash

A star-spanning, Italian space opera, *Starcrash* was the brainchild of writer/director Luigi Cozzi. A long-time SF fan, Cozzi paid on-screen tribute to some of his favorites. The skeletal robot (right) for instance, was designed after the animated skeleton warriors created and used by Ray Harryhausen in *Jason & the Argonauts.*

Starship Invasions

The Earth hangs in the balance, as two warring galactic races bring their conflict here in this throwback to the "B" movies of the fifties. The humor in the film is mostly unintentional . . . witness the robot with the funnel-shaped nose.

PARAMOUNT PRESENTS
THE COLOSSUS OF NEW YORK
STARRING
JOHN BARAGREY
MALA POWERS
OTTO KRUGER
ROBERT HUTTON
ROSS MARTIN
PRODUCED BY
WILLIAM ALLAND
DIRECTED BY
EUGENE LOURIE
SCREENPLAY BY
THELMA SCHNEE
BASED ON A STORY BY WILLIS GOLDBECK

INVASION FROM OUTER SPACE!
"DEVIL GIRL FROM MARS
Produced by Edward J. and Harry Lee Danziger • A Spartan Release

3D
MOON MONSTERS LAUNCH EARTH ATTACK!
ROBOT MONSTER
IN TRU-STEREO 3DIMENSION
with GEORGE NADER
CLAUDIA BARRETT
Produced by AL ZIMBALIST
RELEASED THRU ASTOR PICTURES CORP.

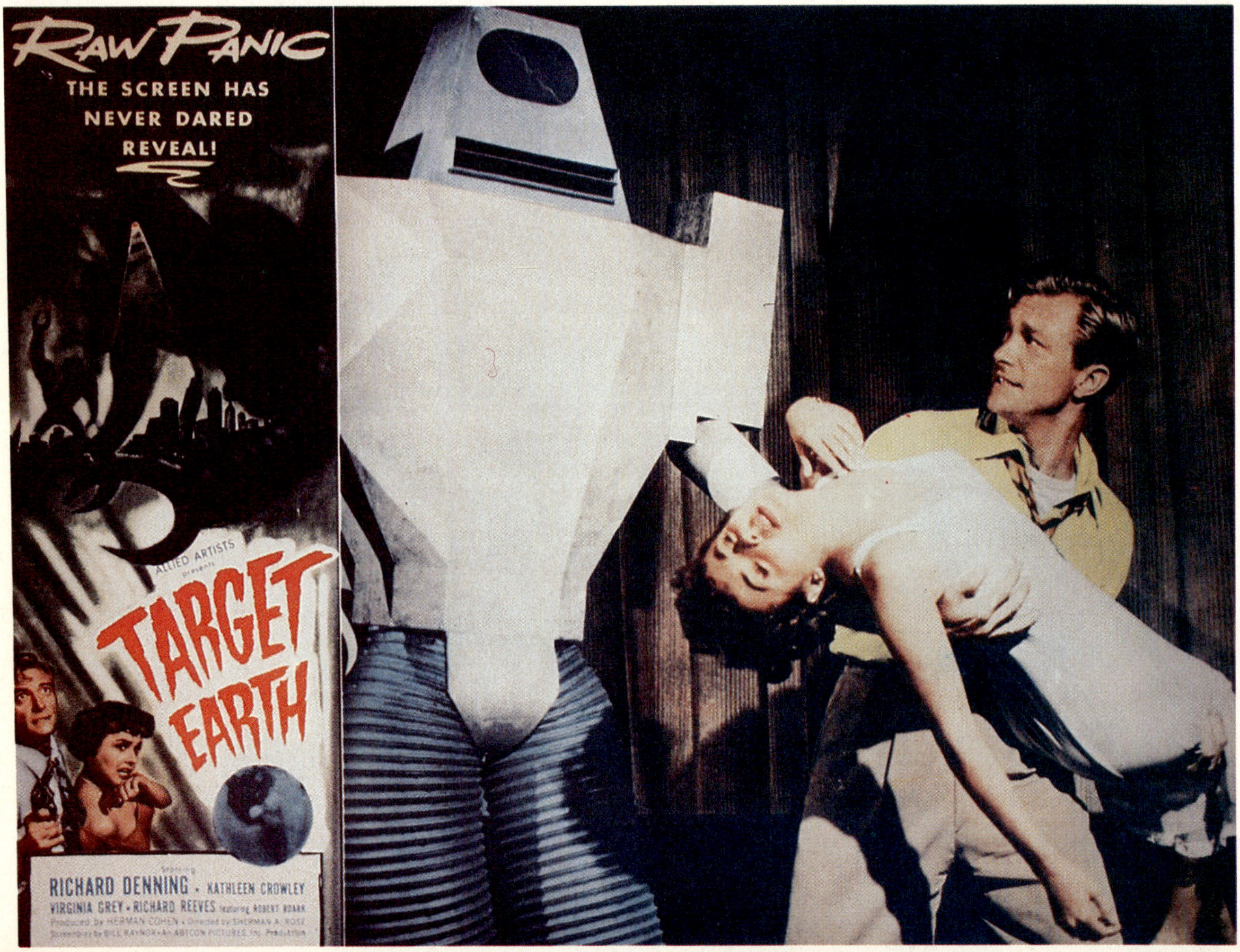
RAW PANIC
THE SCREEN HAS NEVER DARED REVEAL!
ALLIED ARTISTS presents
TARGET EARTH
RICHARD DENNING • KATHLEEN CROWLEY
VIRGINIA GREY • RICHARD REEVES featuring ROBERT ROARK
Produced by HERMAN COHEN • Directed by SHERMAN A. ROSE

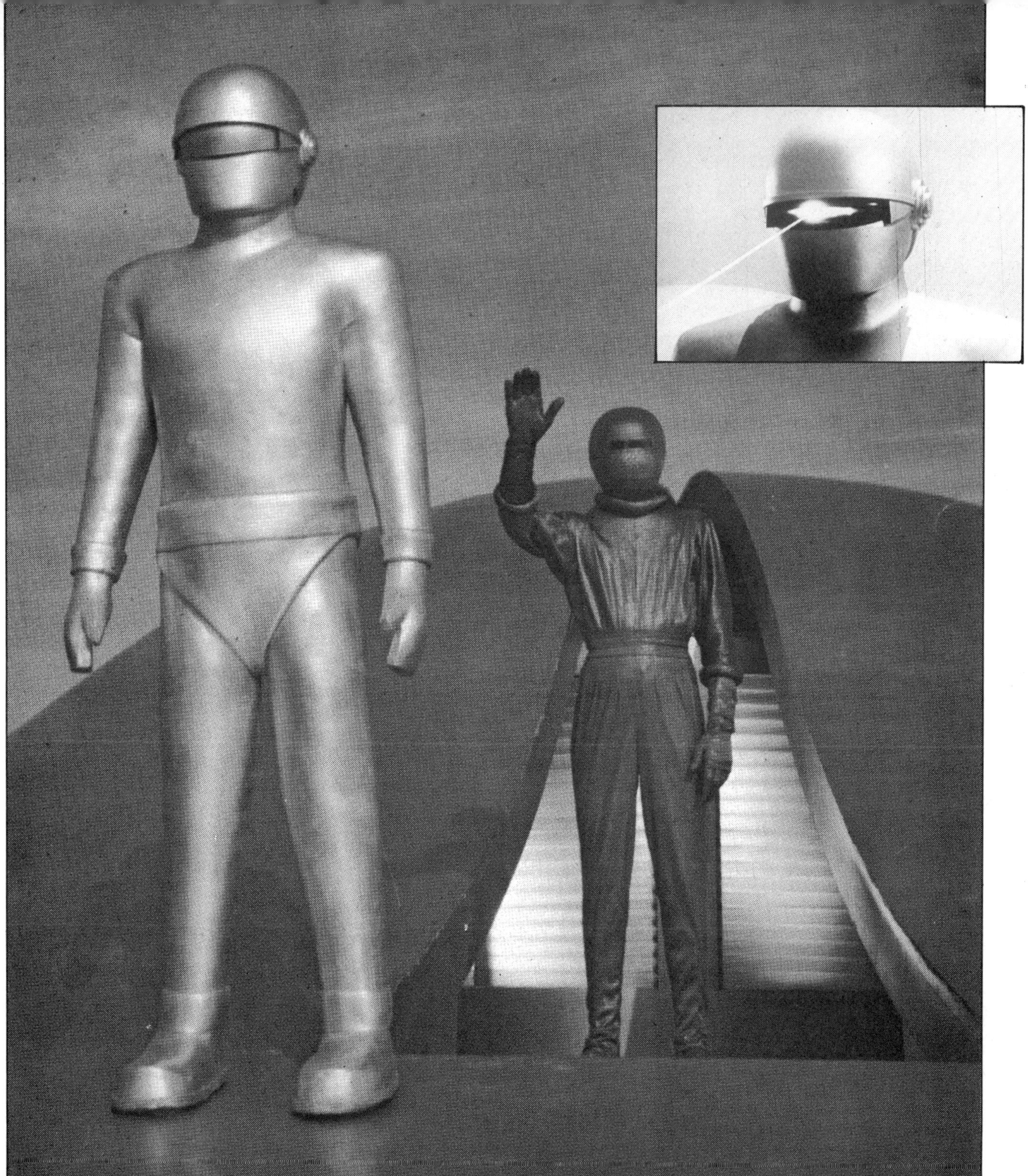

Gort (played by Lock Martin) is an alien robot guardian/policeman, capable of awesome, punitive power in *The Day The Earth Stood Still*, 1957.

MARCH OF THE WOODEN SOLDIERS (aka BABES IN TOYLAND)
MGM, 1934, B&W, 77 min.

Adapted from the Victor Herbert operetta "Babes in Toyland," the film gives Stan Laurel and Oliver Hardy an opportunity to get mixed up with bogeymen and marching robot dolls.
Directors: Gus Meins, Charles Rogers. Producer: Hal Roach. Screenplay: Frank Butler, Nick Grinde. Cinematography: Francis Corby, Art Lloyd. Music: Victor Herbert, Harry Jackson. Supporting Players: Charlotte Henry, Felix Knight, Henry Kleinbach.

MASTER OF THE WORLD
Ariel-Film (Germany), 1934, B&W, 90 min.
This film shows us a world where robots do the dangerous work, with the displaced workers given jobs pushing the buttons that control the robots.
Director: Harry Piel (see also: THE MIRACLE OF TOMORROW). Screenplay: George Muehlen-Schulte. Cast: Walter Janssen, Sybille Schmitz, Walter Franck, S. Schuerenberg.

THE VANISHING SHADOW
Universal, 1934, B&W, serial in 12 parts.

A man whose father was hounded to death by an evil political group seeks revenge, using a large robot, a vest of invisibility and a death ray.
Director: Louis Friedlander (later known as Lew Landers). Screenplay: Het Manheim, Basil Dickey, George Morgan, from a story by Ella O'Neill.
Art Director: Thomas F. O'Neill.
Cast: Onslow Stevens, Ada Ince, Walter Miller, James Dirkin, Richard Cramer,

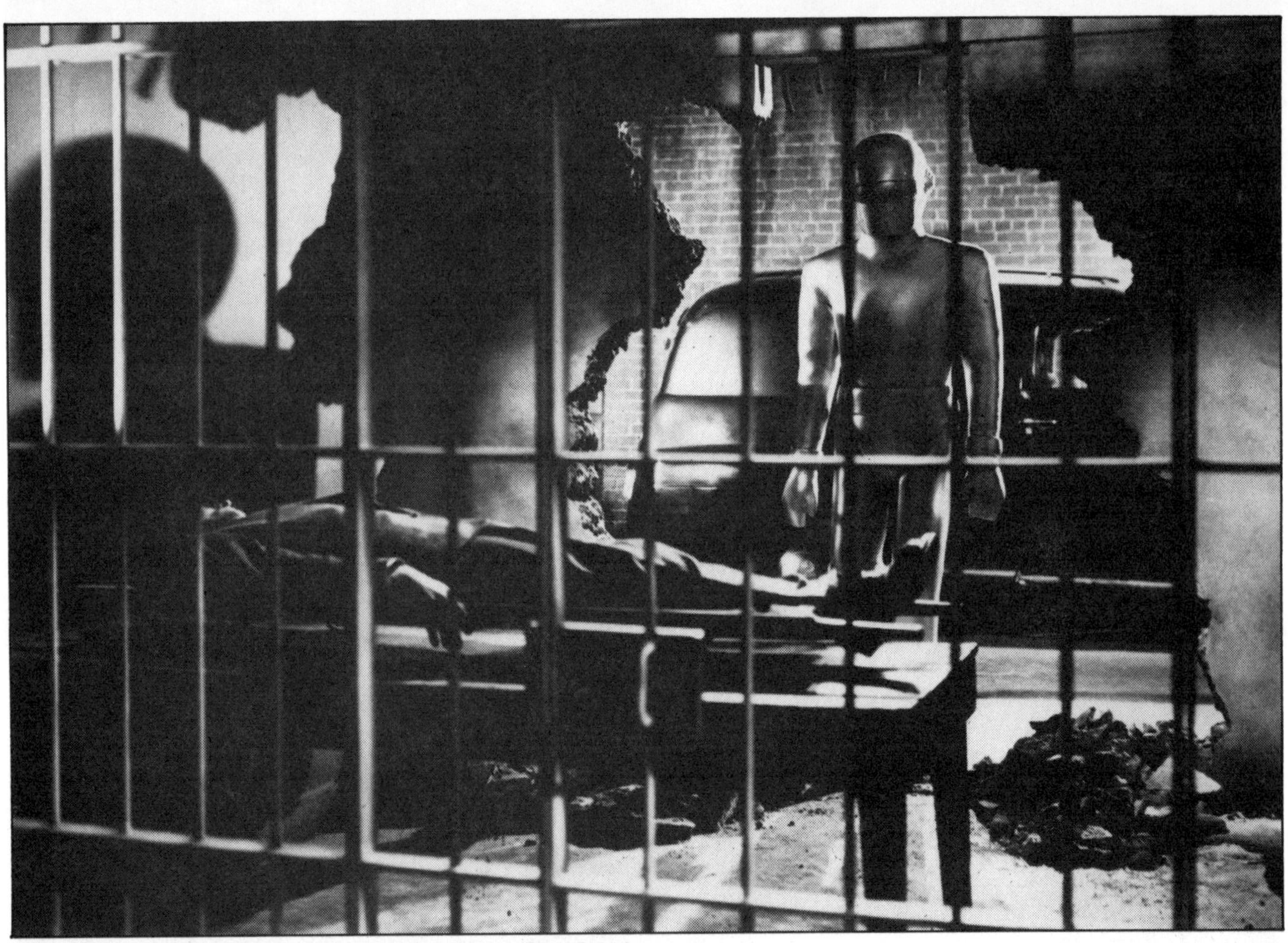

Gort rescues an injured Klaatu from a local police station (above) and scares the wits out of—but doesn't harm—Patricia Neal (bottom), one of the few Earthpeople who befriended Klaatu..

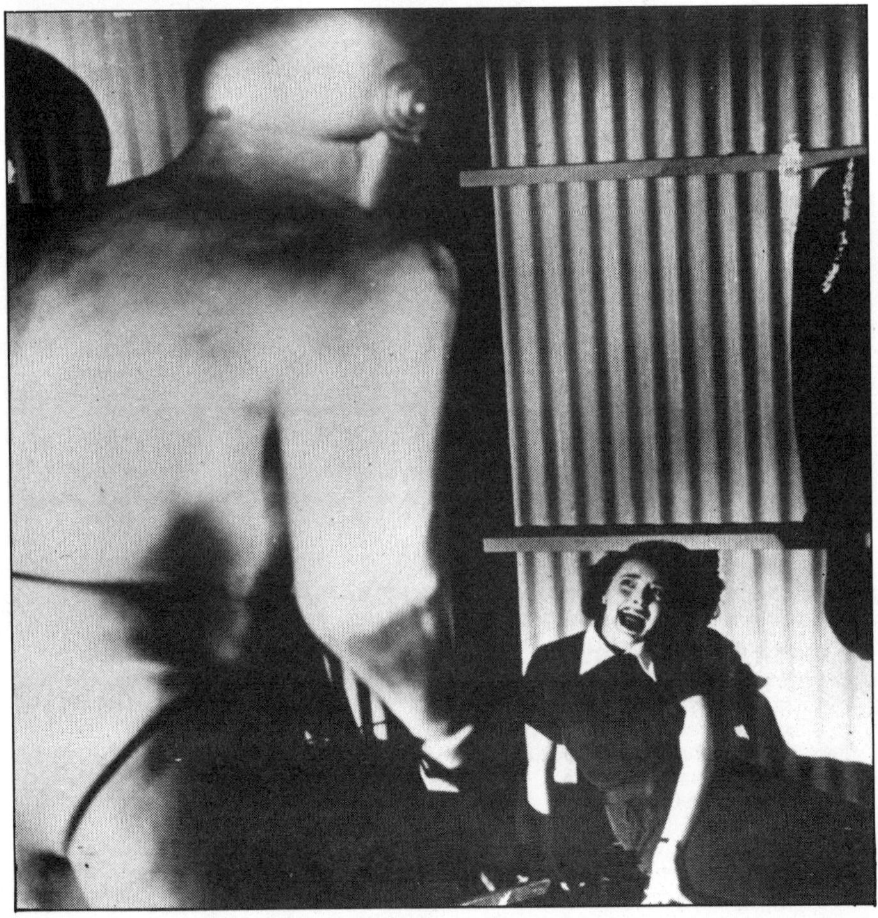

William Desmond, Monte Montague, William Steele, Tom London.

BIRTH OF A ROBOT
(aka THE BIRTH OF THE ROBOT)
Shell Oil (Great Britain), 1934, Color, 7 min.
A film using puppets to show the creation of robots and their capabilities.
Directors: Humphrey Jennings, Len Lye.

THE GOLEM
Metropolis (Czechoslovakia-France), 1935, B&W, 91 min.
This film picks up where the 1920 German film left off. The clay figure of the Golem (Ferdinand Hart), which comes to life when a slip of paper with the words "Revolt is the right of the slave" is placed under its tongue, has been rendered inanimate and placed in the care of Rabbi Jacob (Charles Dorat). The mad Emperor of Prague, Rudolph II (Harry Baur), has reduced the Jews there to slavery. He seizes the clay figure and brings it to the palace. Hoping to use the Golem for his own evil purposes, he imprisons Rabbi Jacob and tortures him to learn the secret of bringing it to life. Rabbi Jacob resists, and his wife succeeds in placing the magic words in the Golem's mouth. It rampages through the palace, avenging the cruel treatment inflicted on the people it was created to protect.
Director: Julien Duvivier. Screenplay: Andre-Paul Antoine. Music: Kumok. Supporting Players: Roger Karl, Gaston Jacquet, Roger Duchesne, Aimos.

THE PHANTOM EMPIRE
Mascot, 1935, B&W, serial in 12 chapters.
While chasing radium thieves across his ranch, Gene Autry falls into a 20,000-foot hole. He lands in Murania, ruled by evil Queen Tika (Dorothy Christy) and her partner in crime, Prime Minister Argo (Wheeler Oakman). These two are planning to conquer the upstairs world with death rays and robots, and imprison Gene and his friends who have followed him down the hole. After being menaced on numerous occasions by the robots, who have flamethrower hands,Gene and his friends escape back to the surface by elevator during a revolt. Murania is finally destroyed by its own death ray during the confusion.
Directors: Otto Brewer, B. Reeves Eason. Screenplay: Wallace MacDonald, Gerald Geraghty, H. Freedman. Supporting Players: Frankie Darrow, Betsy King Ross, Smiley Burnette, Jack Carlyle.

THE TIN MAN
MGM, 1935, B&W, 19 min.
Two women (Patsy Kelly, Thelma Todd), lost in their car, run across a mad scientist and his robot (Cy Slocum). The robot has been ordered by his misogynist master to destroy all women, but suffers from gender confusion and finally attacks the scientist.
Producer: Hal Roach. Director: James Parrott.
Supporting Players: Matthew Betz, Clarence Hummel Wilson.

THE UNDERSEA KINGDOM
Republic, 1936, B&W, serial in 12 chapters.
Earthquakes shaking the North American continent have been traced to the ocean floor, and a submarine equipped with an Earthquake-Counter-Acting Ray has been sent to investigate, commanded by naval officer Crash Corrigan (Ray "Crash" Corrigan). He and his scientific party discover the Lost Continent of Atlantis, an ancient civilization protected from the ocean by a huge dome. A civil war is raging between the White Robes of Sharad (William Farnum), the rightful ruler of Atlantis and the Black Robes of Khan (Monte Blue). Khan's forces possess an atomic-powered disintegration ray and robot warriors called Volkites. He wants to use the submarine's Earthquake Ray to conquer Atlantis and then the whole world, but Corrigan allies himself with the White Robes (did you really not suspect he might?) and restores Sharad to his throne. The film is another of those strange conglomerations of the old and the new, with spears, chariots and bows and arrows mixed in with ray guns, robots and flying machines.
Directors: B. Reeves Eason, Joseph Kane. Screenplay: John Rathmell, Maurice Geraghty. Oliver Drake. Music: Harry Grey. Supporting Players: Loise Wilde, C. Montague Shaw, Smiley Burnette, Lon Chaney, Jr., Lane Chandler, Raymond Hatton.

MECHANICAL HANDY MAN
Winkler Pictures (Universal), 1937, B&W, 7 min.
Oswald the Rabbit, a character originally created by Walt Disney, builds a robot helper in this animated cartoon.
Producer: Walter Lantz. Story: Charles Bowers. Animators: Ted Dubois, Ed Benedict.

BUCK ROGERS
Universal, 1939, B&W, serial in 12 chapters.
Buck (Larry "Buster" Crabbe) and his pal Buddy Wade (Jackie Moran) are overcome by Nirvano gas while piloting a dirigible. They awaken 500 years in the future, in a world ruled by the futuristic hoodlum Killer Kane (Anthony Warde). Kane maintains his control through the use of mind-controlling helmets which turn the wearers into robot-like zombies. He is opposed by Dr. Huer (C. Montague Shaw) and Wilma Deering (Constance Moore), who quickly enlist Buck and Buddy to their cause. Taking a spaceship to Saturn, they put down a rebellion of the sub-human Zuggs and gain the aid of Saturn's ruler, Prince Tallen (Philson Ahn).
Kane is overthrown, Buck is made Air Marshal of Earth, and he and Wilma find happiness together.

The Republic robot strikes again, this time in *Zombies of the Stratosphere,* 1952.

Directors: Ford Beebe, Saul A. Goodkind. Screenplay: Norman S. Hall and Ray Trampe, based on the novels by Philip Francis Nowland and the syndicated comic strip by Phil Nowlan and Dick Calkins. Art Directors: Jack Otterson, Ralph De Lacy. Supporting Players: Jack Mulhall, Guy Usher, William Gould, Henry Brandon, Wheeler Oakman, Kenneth Duncan, Carleton Young, Reed Howes.

THE PHANTOM CREEPS
Universal, 1939, B&W, serial in 12 chapters.
Mad scientist Dr. Zorka (Bela Lugosi) has synthesized a new chemical element which can induce suspended animation. He puts this element into metal disks and creates a mechanical spider which completes the reaction when it contacts the disk; suspended animation results for anyone who has the disk in their clothing. His evil plans (whatever they are) are opposed by American Military Intelligence Agent Captain Bob West (Robert Kent) and his sidekick Jim Daly (Regis Toomey). Zorka fights back using an invisibility belt and a huge sneering robot operated by a radio switchboard, but is destroyed in the end, his plans crushed.
Directors: Ford Beebe, Saul A. Goodkind. Screenplay: George Plympton, Basil Dickey, Mildred Barish.
Supporting Players: Dorothy Arnold, Edward Van Sloan, Eddie Acuff, Edwin Stanley, Lane Chandler.

1940-1949

THE MYSTERIOUS DR. SATAN
Republic, 1940, B&W, serial in 15 parts.
Mad Scientist Dr. Satan (Eduardo Cianelli) plans to conquer the world with an army of robots. (Our old friend the Republic Robot appears once again; you know, the one that looks like someone attacked an ashcan with a pair of tin snips?) Anyway, the bad doctor is having trouble controlling his robots from a distance, so he kidnaps Professor Scott (C. Montague Shaw) and forces him to assist in developing a remote-control device. Dr. Satan's evil plans are opposed by the Copperhead (Robert Wilcox), a masked fighter for justice and truth. The doctor sets a series of traps for the hero, but he escapes them all in time for the next episode. Dr. Satan finally sends his chief robot (Tom Steele) to kill the interfering Copperhead, but the masked hero eludes the metal menace and fights his way into the doctor's hidden sanctuary. During a struggle, the Copperhead slips his mask over Dr. Satan's head; confused, the robot grabs the doctor instead of the hero and the two of them plummet out a window many stories high.
Directors: William Witney, John English. Screenplay: Franklyn Adreon, Ronald Davidson, Norman S. Hall, Joseph Poland, Sol Shor. Music: Cy Feuer. Supporting Players: William Newell, Ella Neal, Dorothy Herbert, Charles Trowbridge, Jack Mulhall.

Publicity art shows Ro-Man taking last available woman on Earth, in low-budget atomic nightmare, *Robot Monster*, 1953.

ROBOT WRECKS
MGM, 1941, B&W, 11 min.
The "Our Gang" kids build a robot; you can fill in the rest.
Director: Edward Cahn. Screenplay: Hal Law, Robert McGowan. Cast: George "Spanky" McFarland, Darla Hood, Mickey Gubitosi (Robert Blake), Billy Laughland, Billy Bletcher, Billie Thomas, Margaret Bert.

DEAD OF NIGHT
Universal, 1945, B&W, 100 min.
An episodic film built around a recurring dream of Walter Craig (Mervyn Johns) in which he discusses psychic phenomena with various people. One episode concerns a mad ventriloquist, Maxwell Frere (Michael Redgrave), who becomes convinced his dummy (Johnny Maguire) is alive and wishes to break its partnership with him and run off with another ventriloquist. Frere shoots his rival, and is sent to prison. While there, he is given back his dummy, since only through it can he speak; he attacks and destroys it, and is himself destroyed.
Directors: Alfredo Cavalcanti (director of ventriloquist episode), Charles Crichton, Basil Dearden, Robert Hamer. Screenplay: John Baines, Angus MacPhail. Music: George Auric. Supporting Players: Roland Culver, Googie Winters, Sally Ann Howes, Miles Malleson.

THE MONSTER AND THE APE
Darmour, 1945, B&W, serial in 15 parts.
Professor Arnold (Ralph Morgan) has developed a new metal, Metalogen, and constructed a robot from it. A criminal conspiracy led by Professor Ernst (George Macready) has stolen the robot, using a trained ape named Thor (Ray "Crash" Corrigan). After many misadventures the robot is recovered by dashing Ken Morgan (Robert Lowery), and Professor Ernst is killed, ending the conspiracy.
Director: Howard Bretherton. Screenplay: Sherman Lowe, Royal K. Cole. Music: Lee Zahler. Supporting Players: Carole Mathews, Willie Best, Jack Ingram, Anthony Warde, Ted Mapes, Eddie Parker, Stanley Price.

THE PERFECT WOMAN
Two Cities (Great Britain), 1949, B&W, 89 min.

Two eccentric men (Stanley Holloway, Nigel Patrick) take on the job of road-testing a female robot invented by Professor Belmon (Miles Malleson). As a joke, the professor's niece (Patricia Roc) substitutes herself for the robot. Amazed at how life-like the robot is, the two men take it to the bridal suite of a hotel.
Director: Bernard Knowles. Screenplay: Bernard Knowles, George Black, from a play by Wallace Geoffrey and Basil John Mitchell. Supporting Players: Irene Handl, Pamela Devis.

THE MAGIC SWORD

Zvezdafilm (Yugoslavia), 1949, B&W, 97 min.

Bash-Chelic (Milivoye Zhivanovich), a steel robot monster reigning over the empire as a tyrant, lusts after the maiden, Vida (Vera Ilich-Djukich). Her true love, Neboyshe (Rade Markovich), sets out to find the magic sword which can destroy the robot. After many adventures he succeeds in finding it, kills Bash-Chelic, and finds happiness with Vida.
Writer-Director: Voislav Nanovich. Music: Kresmir Baranovich. Supporting Players: Mihailo Pazskalyevich, Ljubisha Yovanovich, Marko Marinkovich, Zora Zlatovich.

1950-1959

CAPTAIN VIDEO

Columbia, 1951, B&W, serial in 15 chapters.

Captain Video (Judd Holdren) and the Video Ranger (Larry Stewart) are headed for the planet Atoma in their spaceship *The Galaxy* to stop an alliance between the evil Dr. Tobor (George Eldredge) and the ruler of Atoma, Vultura (Gene Roth). Vultura fails to kill them with a remote-controlled comet upon their arrival, and they spend the remainder of the serial battling the forces of Vultura and Tobor, which include a number of robots left over from *The Phantom Empire* of 1935. Vultura is finally killed by his own death ray, and Captain Video and the Ranger return to Earth and their secret headquarters.
Directors: Spencer Gordon Bennet, Wallace A. Grissell. Screenplay: Royal K. Cole, Sherman L. Lowe and Joseph F. Poland based on characters from the TV series *Captain Video and His Video Rangers.* Music: Mischa Bakaleinikoff. Special Effects: Jack Erickson. Supporting Players: Don C. Harvey, William Fawcett, Jack Ingram, I. Stanford Jolley, Skelton Knaggs, Jimmy Stark.

THE DAY THE EARTH STOOD STILL

20th Century-Fox, 1951, B&W, 92 min.

One of the classic SF films. Klaatu (Michael Rennie) and Gort (Lock Martin), his robot traveling companion, arrive in Washington, D.C., in a flying saucer. He is the emissary of a planetary federation which has been observing the Earth for many years, and he has brought a warning: Earth must cease its atomic testing and refrain from exporting its aggressive, war-like behavior into space

Robby serves as motive power for this landcraft in *Forbidden Planet,* 1956.

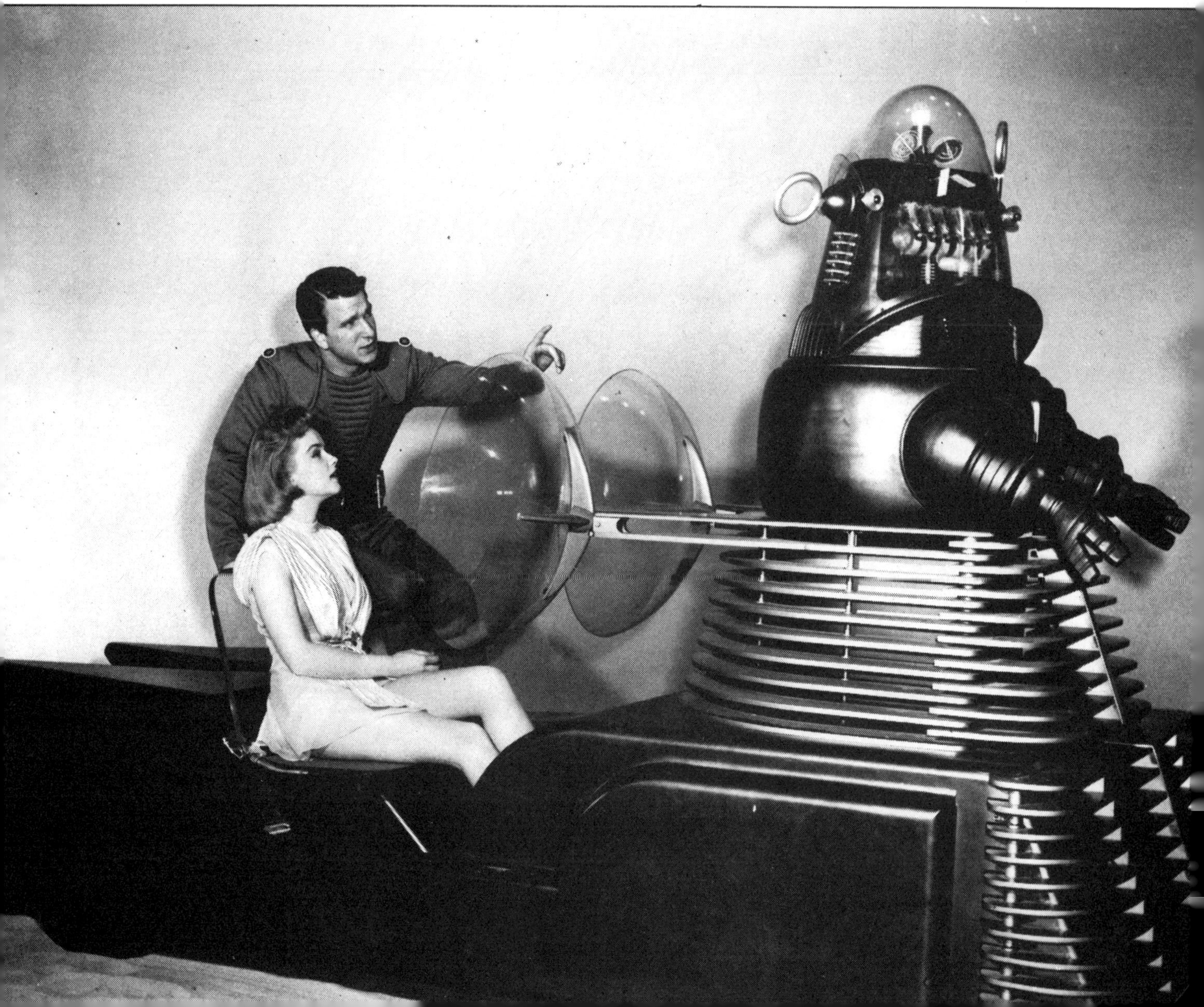

or the planet will be destroyed. To emphasize the power he holds, Klaatu shuts down the Earth's electrical power for 30 minutes. Taking an assumed name, Klaatu mingles with the Earth people, making friends and allies of Professor Barnhardt (Sam Jaffe) and Helen Benson (Patricia Neal) and her son Billy (Billy Gray). Sadly, he is thought to be a hoax by the nations of the world, and is gunned down and killed by the military when he returns to the saucer. Helen activates Gort, who destroys the soldiers and their weapons and brings Klaatu back to life using the awesome technology of the saucer. As he prepares to depart, Klaatu repeats his warning: learn to live in peace or be destroyed by police robots like Gort, to whom all authority has been entrusted. There are a number of religious parallels in the film: the name Klaatu takes is Mr. Carpenter; he has come to save the Earth from its sins; he is killed, resurrected, and ascends to heaven. Gort, the 12-foot robot, was built around seven-foot Lock Martin, a former doorman at Grauman's (now Mann's) Chinese Theatre in Hollywood. There were two costumes, one with an opening in the back for front-filmed scenes, the other with an opening in the front for shots from the rear. There was also a fiberglass replica made for shots when it was not necessary for Gort to move.

Director: Robert Wise. Screenplay: Edmund H. North, from the story "Farewell to the Master" by Harry Bates. Art Directors: Lyle Wheeler, Addison Hehr. Music: Bernard Herrman. Special Effects: Fred Sersen. Supporting Players: Hugh Marlowe, Frances Bavier, Drew Pearson, Frank Conroy, Carleton Young, Edith Evanson, John Brown, Glenn Hardy, House Peters.

ZOMBIES OF THE STRATOSPHERE

Republic 1952, A 12-episode seial, 24 reels, 384 min., B&W

Judd Holdren returned as rocket-belt-wearing Commando Cody in this SF adventure yarn. Three Martian meanies —Marex, Roth and Narab (Lane Bradford, Craig Kelly and Leonard Nimoy)—attempt to gain control of Earth by enlisting the aid of a demented Earth scientist and his robot (a holdover from the Republic serials of the *Mysterious Dr. Satan* era). The Martians plan to construct a bomb which will blast Earth out of its orbit. They don't.

Director: Fred C. Brannon. Screenplay: Ronald Davidson. Music: Stanley Wilson. Special Effects: Howard and Theodore Lydecker.

Supporting players: Aline Towne, Wilson Wood, Johnny Crawford, Dale Van Sicke.

THE EMPEROR AND THE GOLEM

Artkino (Czechoslovakia), 1955, Color, 95 min.

A comic re-telling of the encounter between the Golem and mad Emperor Rudolph II of Prague (Jan Werich), which was treated seriously in the film *The Golem* (1935). Rudolph wants to make the Golem his friend, but he is too addled to keep his mind on it for very long. Through a series of disasters, young Matthew the Baker (also played by Werich) becomes Emperor, and puts the Golem to work as a bread-baking powerhouse.

Director: Martin Fric. Screenplay: Jan Werich, Jiri Brdecka. Music: Julius Kalas. Supporting Players: Frantisek Filiposky, Jiri Plashy, Zdenek Stepanek, Natasa Gollova, Marie Vasova.

TALES OF HOFFMAN

London Films (Great Britain), 1951, Color, 138 min.

In what is basically a filmed version of the opera by Jacques Offenbach, Hoffman (Robert Rounseville) tells the customers in a tavern of his three loves; one of them is Olympia (Moira Shearer), a life-like dancing robot doll.

Writers-Producers-Directors: Michael Powell, Emeric Pressburger.

Music: Jacques Offenbach.

Choreography: Frederick Ashton.

Robby the Robot shows his versatility in *Forbidden Planet*. Below: Robby plays house servant. Right: Robby synthesizes gallons of Bourbon for an appreciative ship's cook (Earl Holliman).

Supporting Players: Robert Helpmann, Pamela Brown, Leonide Massine, Frederick Ashton, Ann Ayars.

ALRAUNE (aka UNNATURAL)

Styria-Carlton (Germany), 1952, B&W, 92 min.

Yet another version of the novel by Ewer, with Hildegard Knef (known in the U.S. as Neff) as Alraune and Erich Von Stroheim as her creator, Ted Brinken. Alraune goes her usual way—every man she comes into contact with dies. But a subplot has been added about Alraune using her mystical powers to find a medicinal spring which Brinken uses to swindle the public until caught by the health authorities. She falls in love with a good man (Karl Bohm) finally, but Brinken, in a jealous incestuous rage, shoots her and goes to the gallows, the final victim of her fatal charm.
Director: Arthur Maria Rabenalt. Screenplay: Fritz Rotter. Music: Werner Heymann. Supporting Players: Julia Koschka, Trude Hesterberg.

ROBOT MONSTER

Astor, 1953, B&W, 63 min., filmed in 3-D.

Ro-Man, a shaggy, ape-like robot alien, wages murder and mayhem with his ray gun, killing all life on Earth except for six people. Just as he is about to kill these last six, literally bringing about the end of the world, we discover it has all been a dream by a small boy who has seen too many films like this one. This film is a perfect example of how budget problems influence content: the robot looks like an ape because the producers could only afford to rent a gorilla costume, lacking the money to build a robot.
Director: Phil Tucker. Screenplay: Wyatt Ordung. Cinematography: Jack Greenhalgh. Special Effects: Jack Rabin, David Commons. Music: Elmer Bernstein. Cast: George Nader, Claudia Barrett, Selena Royle, Gregory Moffett, John Mylong, Pamela Paulson.

THE TWONKY

United Artists, 1953, B&W, 72 min.

An alien robot from the future accidently invades a television set belonging to Terry (Hans Conried), and once settled in proceeds to take over his life. The Twonky wishes only to serve, whether Terry likes it or not. It does his housework, but prevents him from doing anything *he* would like to do. It also walks, talks, and defends itself through hypnotism. The movie attempts to make the point that machines can free man from physical tasks, but can also turn mankind into automatons. Belatedly realizing this, Terry destroys the Twonky.
Producer-Director: Arch Oboler. Screenplay: Arch Oboler from a story by Henry Kuttner. Music: Jack Meakin. Supporting players: Gloria Blondell, Billy Lynn, Janet Warren, Ed Max.

Billy Chapin with remote-controlled robot and friend, from *Tobor the Great*, 1954.

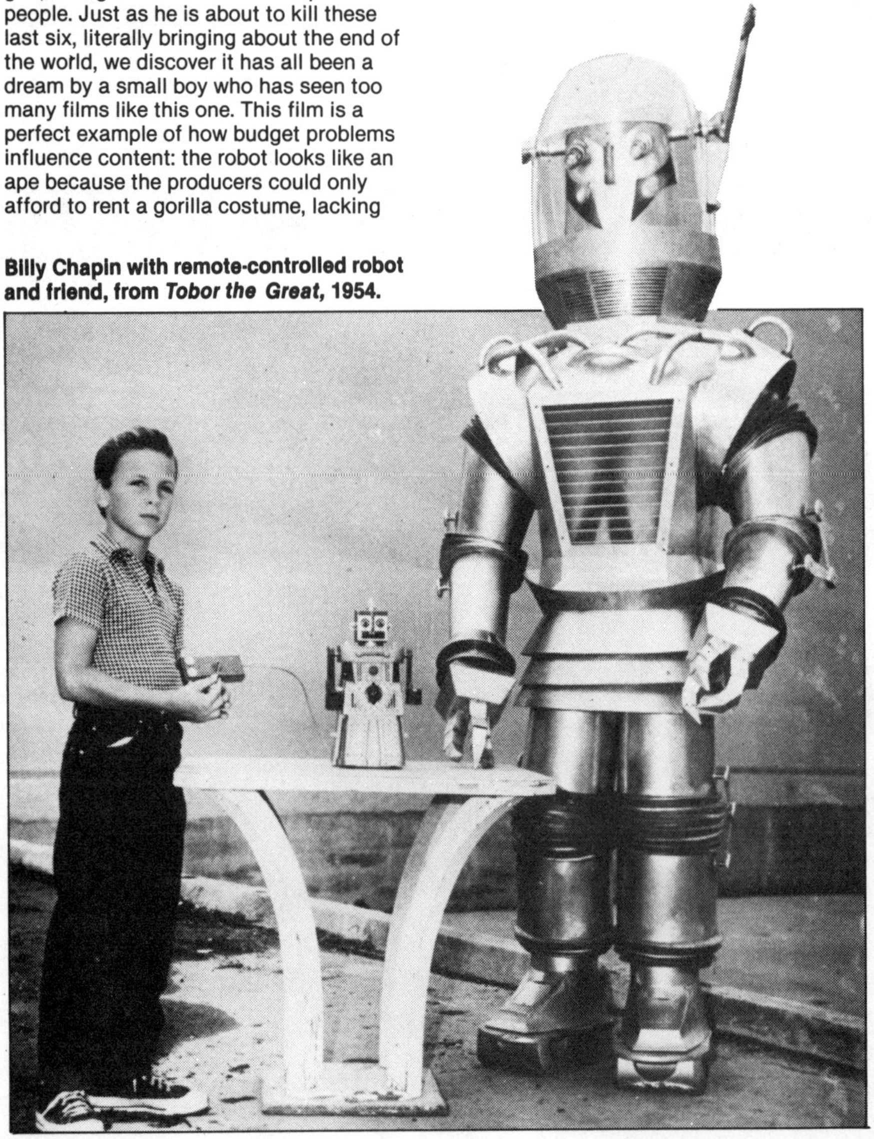

Tobor puts adventure back into the Western custom of shaking hands.

DEVIL GIRL FROM MARS

British Lion (Great Britain), 1954, B&W, 76 min.

What more is there to say after you've heard the title? Nyah the Devil Girl (Patricia Laffan) and her robot Chani arrive in Scotland by flying saucer in this low-budget British knock-off of *The Day the Earth Stood Still*. At first it appears her mission is a peaceful one, but it is soon revealed that her true purpose is to kidnap virile males for breeding stock for her depopulated planet. The Highlanders resist this nefarious plot vigorously; perhaps she would have done better to land in Italy. Becoming testy and violent over this resistance, she and Chani wreak havoc on the heather until they are finally destroyed.
Director: David MacDonald. Screenplay: John C. Mather and James Eastwood, adapted from their play. Music: Edwin Astley. Special Effects: Jack Whitehead. Supporting Players: Hugh McDermott, Hazel Court, Adrienne Corri, Peter Reynolds, Joseph Tomelty, Sophie Stewart.

GOG (aka GOG, THE KILLER)

United Artists, 1954, Color, 85 min., filmed in 3-D.

Gog and Magog are two robots working at a space station project in New Mexico under the direction of Novac, a giant computer. An enemy plane attempts to sabotage the project by beaming a destructive ray at Novac, which causes the two robots to go berserk: they begin destroying the station and try to kill the scientists are work there. The rampaging robots are finally destroyed by Dr. Van Ness (Herbert Marshall) and David Sheppard (Richard Egan). The two robots are distinctly non-humanoid in appearance: they possess multiple arms designed for various purposes and run on tank-type treads.

Director: Herbert L. Strock. Producer: Ivan Tors. Screenplay: Tom Taggart. Music: Harry Sukman. Special Effects: Harry Redmond, Jr. Supporting Players: Constance Dowling, Philip Van Zandt, William Schallert.

TARGET EARTH

Allied Artists, 1954, B&W, 75 min.

Nora (Kathleen Crowley) awakes to find the city deserted, occupied by robot invaders from Venus. She meets up with Frank (Richard Denning) and some other stragglers and learns they are the only survivors of the attack, made by the robots using death ray projectors located in their heads. The robots are finally destroyed by a scientist (Whit Bissell), but only Frank and Nora survive.
Director: Sherman A. Rose. Screenplay: William Raynor, from the story "Deadly City" by Ivar Jorgenson. Special Effects: Dave Koehler. Music: Paul Dunlap. Supporting Players: Virginia Grey, Richard Reeves, Arthur Space, House Peters, Jr.

TOBOR THE GREAT

Republic, 1954, B&W, 77 min.

An inventor has developed Tobor, an experimental robot with emotions who can receive telepathic impulses. This comes in very handy when the inventor and his grandson are kidnapped by spies who want to put the scientist to work for their government: Tobor is summoned, and comes charging to the rescue. One of the few early robot heroes.

Director: Lee Sholem. Screenplay: Phillip MacDonald, Richard Goldstone. Special Effects: Howard Lydecker, Theodore Lydecker. Music: Howard Jackson. Cast: Charles Drake, Karin Booth, Billy Chapin, Taylor Holmes, Lyle Talbot, William Schallert, Henry Kulky, Robert Shayne.

ROBOT RABBIT

Warner Bros. Cartoon, 1954, Color, 7 min.

Bugs Bunny (voice: Mel Blanc) has trouble getting past the robot guard of a carrot patch—probably programmed by E. Fudd. Director: I. Freling. Story: Warren Foster. Animators: Virgil Ross, Arthur Davis, Manuel Perez, Ken Chappin. Music: Carl Stalling.

THE BOWERY BOYS MEET THE MONSTERS

Allied Artists, 1954, B&W, 65 min.

The Bowery Boys, led by Leo Gorcey and Huntz Hall, run into problems when they try to rent a vacant lot owned by a family of madmen: they are set upon by a robot, a gorilla and a man-eating tree. Huntz Hall is finally turned into a monster, but no one notices.
Director: Edward Bernds. Screenplay: Edward Bernds, Ellwood Ullman. Special Effects: Augie Lohman. Makeup: Edward Polo. Music: Marlin Skiles.
Supporting Players: Ellen Corby, John Dehner, Laura Mason, Lloyd Corrigan.

Huntz Hall has fun as *The Bowery Boys Meet the Monsters*, 1954.

Devil girl Nyah and Chani the robot in *Devil Girls from Mars,* 1954.

THE FLYING SAUCERS
Clasa-Mohme (Mexico), 1955, B&W, length unknown.
A Mexican plumber builds a flying saucer; anyone who has ever had occasion to use a toilet in Tijuana will find this easy to believe. When it crashes, he and his companion are mistaken for Martian robots.
Director: Julian Soler. Screenplay: Carlos Orellana, Pedro de Urdimalas. Special Effects: Armando Stahl. Music: Manuel Esperon.
Cast: "Resortes," Evangelina Elizondo, Andres Solar, Jose Venegas, Amalia Aguilar, Pedro de Urdimalas.

EARTH VS. THE FLYING SAUCERS
Columbia, 1956, B&W, 83 min.
Flying saucers have been a source of fascination since the first reported sighting by a pilot in the state of Washington in June, 1947, and they are the central feature of interest in this film, beautifully depicted in stop-motion animation by Ray Harryhausen and Russ Kelley. A small fleet of saucers descends upon a government installation and destroys it with heat rays. The robot pilots of the saucers then summon a scientist, Dr. Russell Marvin (Hugh Marlowe), and inform him they have left their dying world to resettle on Earth; if the world governments will not cooperate, the aliens will seize the planet by force. Dr. Marvin responds by inventing a weapon which will unbalance the saucers' gyroscopic stability and knock them out of the sky. He succeeds in destroying the invaders, though not before a couple of the saucers have destroyed the Washington Monument and the Capitol Building. It is discovered that the aliens are not robots, but extremely old and fragile beings using robotic bodies as a protective skin.
Director: Fred F. Sears. Screenplay: George Worthington Tates, Raymond T. Marcus. Story adapted by Curt Siodmak from the book "Flying Saucers From Outer Space" by Donald E. Keyhoe. Music Mischa Bakaleinikoff. Special Effects: Ray Harryhausen, Russ Kelley.
Supporting Players: Joan Taylor, Morris Ankrum, Donald Curtis, Tom Browne Henry, Grandon Rhodes, Larry Blake.

FORBIDDEN PLANET
MGM, 1956, Color, 98 min.
Academy-Award-nominated special effects provide much of the excitement in this film, loosely based on William Shakespeare's play, "The Tempest." Searching for evidence of the fate of a space colony settled on the planet Altair-4 25 years earlier, Relief Saucer C57D lands on the desolate planet in the year 2200 A.D. The crew (Leslie Nielson as Commander Adams, Warren Stevens as Doctor Ostrow, Jack Kelly as Lieutenant Farman, Richard Anderson as Chief Quinn, and Earl Holliman as the cook) is met by Dr. Morbius (Walter Pidgeon) and his daughter, Altaira (Anne Francis), the only two survivors of the colony. They are accompanied by Robby the Robot (voice: Marvin Miller), contructed by Morbius with knowledge he had gleaned from studying the scientific records of the Krel—a long-dead civilization which once flourished on the planet. Robby is totally sophisticated (as robots go). He is constrained by his programming from harming human life and is dedicated to caring for Altaira. Robby also possesses the ability to manufacture *anything*, once given a sample for analysis. Morbius explains that all the other members of the space colony except himself and his wife and daughter were murdered shortly after their arrival on Altair by a mysterious force. Though less than pleased by the arrival of the saucer and its crew, he shows the officers the underground laboratories and machinery of the Krel, including a device to boost intelligence which he used shortly after his arrival on the planet. Altaira becomes romantically involved with Commander Adams, and soon afterwards the mysterious force begins to strike again, butchering several members of the crew. Doctor Ostrow attempts to use the Krel mind-boost device, and though the shock of it kills him, he discovers the hidden secret of the murderous force: one of the huge Krel machines is a device capable of animating energy into matter by thought alone. Connected to the brain of every living Krel, it was intended to free them from all manual labor, thereby giving them more time for scientific pursuits. But the Krel failed to consider the repressed hatreds and jealousies existing in the subconscious mind: these were also given concrete form in the shape of murdering monsters, and the Krel civilization was wiped out in one mad night of murder. Morbius, by using the Krel mind-boost device, became capable of activating the energy shaping machine, and killed his fellow colonists with his unconscious thoughts. Realizing this, Morbius

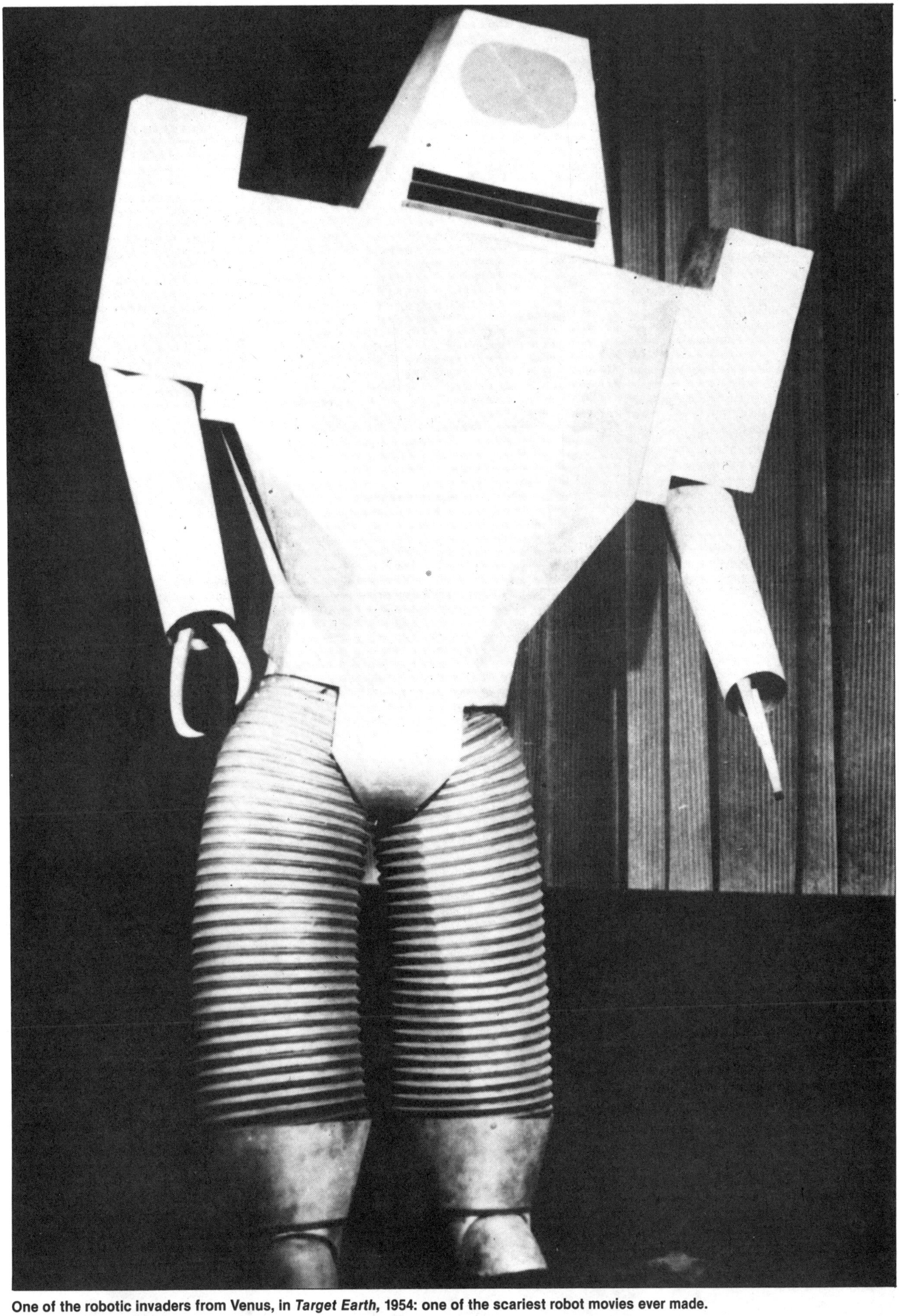

One of the robotic invaders from Venus, in *Target Earth,* 1954: one of the scariest robot movies ever made.

sacrifices his life and destroys the planet, allowing his daughter, Robby and the surviving crew members of Saucer C57D to escape. Robby, the full extent of whose Krel-designed powers is never truly explored, pilots the craft.
Director: Fred M. Wilcox. Producer: Nicholas Nayfack. Screenplay: Cyril Hume, from a story by Irving Block and Allen Adler. Art Directors: Arthur Lonergan, Cedric Gibbons. Electronic Tonalities: Bebe Barron, Louis Barron. Special Effects: A. Arnold Gillespie, Warren Newcombe, Irving G. Ries. Animation: Joshua Meador. Robby the Robot designed by Bob Kinoshita, from sketches by A. Arnold Gillespie. Supporting Players: James Drury, George Wallace, Bob Dix, Jimmy Thompson, Harry Harvey, Roger McGee, Peter Miller, Morgan Jones, Richard Grant.

Above: Richard Egan explains the workings of *GOG* (1954) before it goes berserk (below).

THE INVISIBLE BOY

MGM, 1957, B&W, 90 min.

Robby the Robot's second appearance in films, but this is most definitely *not* a sequel to *Forbidden Planet*, though the two films also share the same producer. Robby is the property of Dr. Merrinoe (Philip Abbot), who has also developed a super computer currently used in the U.S. space program. Robby becomes friends with Dr. Merrinoe's son Timmie (Richard Eyer), and makes him invisible so that they can play secretly. Aliens from outer space take control of Robby and he is turned evil—using Timmie as a hostage to gain access to secret information in the computer. The control is finally broken, and the aliens defeated.
Producer: Nicholas Nayfack. Director: Herman Hoffman. Screenplay: Cyril Hume, based on a story by Edmund Cooper. Special Effects: Jack Rabin, Irving Block, Louis Dewitt. Music: Les Baxter.

KRONOS

20th Century-Fox, 1957, B&W, 78 min.

A giant fireball has crashed into the ocean off Mexico after streaking through the Earth's atmosphere. What emerges on the beach the next day is a huge tripodic robot, sent by the liquid-energy creatures of another world to absorb all of the Earth's energy for transmission back to their own planet. It begins a relentless march across the country, wreaking havoc and destruction as it makes its way from one source of power to another; even atomic bombs are merely absorbed by Kronos for their energy. Finally, scientists Les (Jeff Morrow) and Culver (George O'Hanlon) destroy Kronos by having it dusted with atomic particles, which causes a massive short-circuit and releases all the stored energy in a fiery blast. The robot, named for the warrior father of Zeus, King of the Gods, was a combination of cartoon animation and sectional constructions of the head and legs.
Producer-Director: Kurt Neumann. Screenplay: Lawrence L. Goldman. Music: Paul Sawtell. Special Effects: Jack Rabin, Irving Block, Louis DeWitt.
Supporting Players: Barbara Lawrence, John Emery, Morris Ankrum, Kenneth Alton, John Parrish.

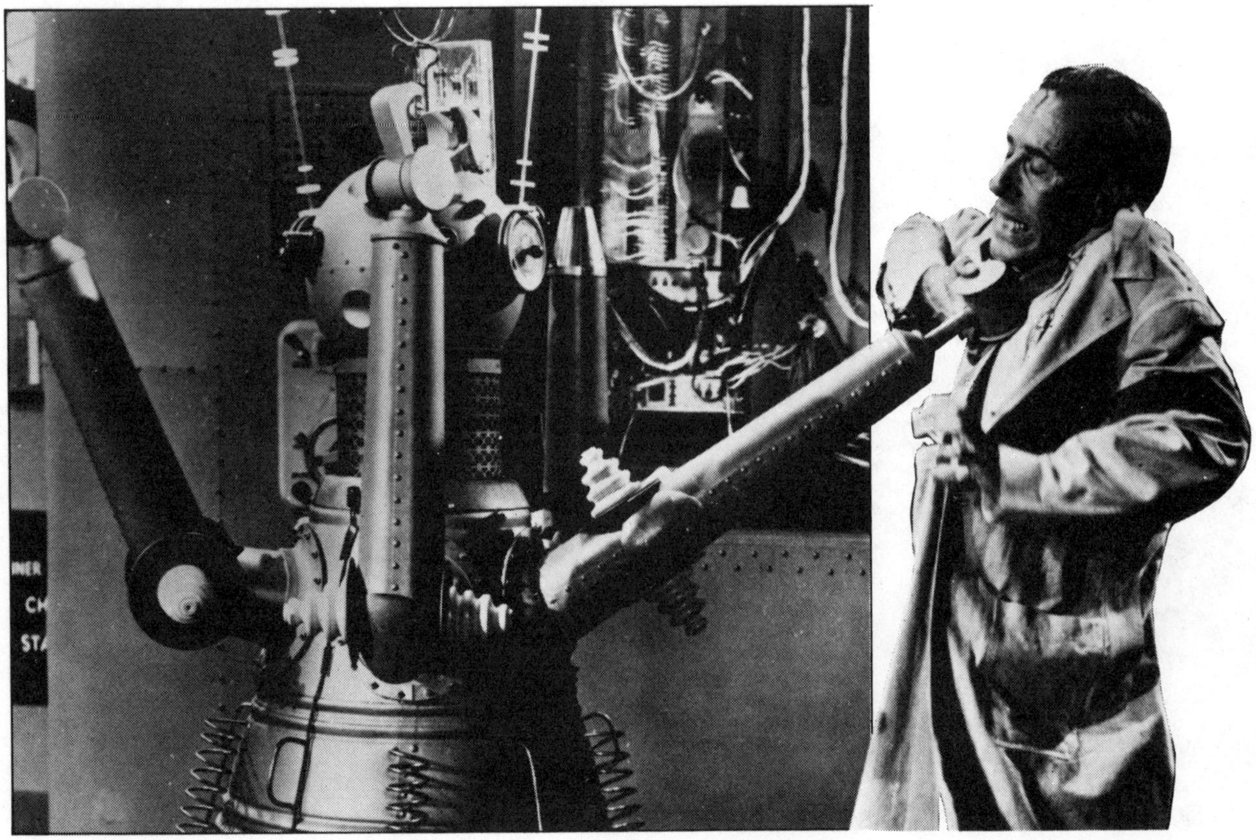

Richard Eyer befriends Robby in *The Invisible Boy*, 1957. In the background is the American "super-computer."

THE MYSTERIANS (aka CHIKYU BOEIGUN, aka EARTH DEFENSE FORCE)

Toho (Japan), 1957, Color, 89 min.

The home planet of The Mysterians has been destroyed in an atomic explosion, so an advance party has been sent to seize the Earth. They intend to enslave all the Earth men and breed with the women. To accomplish this they turn loose their giant robot, fifty times the size of a human, which can shoot death rays from its eyes. The robot is soon destroying much of Japan, but is itself destroyed when it falls into a river and short-circuits. This demoralizes the Mysterians, who have meanwhile been chasing humans all over the countryside in their flying saucers, and plans for the invasion are ended; perhaps they only had one robot.

Director: Inoshiro Honda. Screenplay: Takeshi Kimura. Music: Akira Ifukube. Special Effects: Eiji Tsuburaya.
Cast: Kenji Sahara, Yumi Shirakawa, Akihiko Hirata, Momoko Kochi, Takashi Shimura.

Robby acts under the control of evil aliens in *The Invisible Boy.*

TWENTY FACES

Toei (Japan), 1957, B&W, 113 min.

An escaped convict who can disguise himself at will with any one of twenty faces uses robot beetles and robot monsters in his fatal attempt to steal some atomic material.

Director: Hideo Sekigawa. Screenplay: Tadashi Ogawa. Cinematography: Hiroshi Fukushima.
Cast: Eiji Okada, Jun Usami, Takashi Kanda, Masako Nakamura, Mitsue Komiya.

THE COLOSSUS OF NEW YORK

Paramount, 1958, B&W, 70 min.

When Dr. Jeremy Spensser (Ross Martin) is killed in an accident, his father, Dr. William Spensser (Otto Kreuger), transplants Jeremy's brain into a towering robot body. Lacking a soul, this new creation runs amok through New York City shooting death rays from its eyes until brought under control by Jeremy's

The gigantic, alien, energy-stealing robotic machine known as *Kronos* thunders its way across the Mexican countryside.

young son, Billy (Charles Herbert). Ross Martin could have almost phoned in his performance, since he is only seen in the first few minutes of the film, and another actor (Ed Wolff) wore the Colossus body; a situation similar to James Earl Jones being the voice for David Prowse's portrayal of "Darth Vader" in *Star Wars*.
Director: Eugene Lourie. Screenplay: Thelma Schnee, based on a story by Willis Goldbeck. Music: Van Cleave.
Special Effects: John P. Fulton.
Supporting Players: Mala Powers, John Baragrey, Robert Hutton.

HOW TO MAKE A MONSTER
AIP, 1958, B&W with a Color sequence, 75 min.
Veteran movie makeup artist Pete Drummond (Robert H. Harris) has been fired by the New York executives who have taken over the Hollywood studio where he works. They have also stopped production of the latest horror film on which he was working. Unknown to them, Drummond has developed a new makeup which, when applied externally, produces the same effect as a prefrontal lobotomy. While this would go far in explaining much of the acting seen in "B" robot films, Drummond's purpose is revenge: he makes up several actors as Frankenstein monsters and werewolves and sends them all out to kill the executives.
Director: Herbert L. Strock. Screenplay: Kenneth Langtry, Herman Cohen. Music: Paul Dunlap.
Supporting Players: Paul Brinegar, Gary Conway, Gary Clarke, Malcom Atterbury, Dennis Cross.

1960-1969

THE ROBOT VS. THE AZTEC MUMMY (aka EL ROBOT HUMANO)
K. Gordon Murray Productions (Mexico), 1960, B&W, 65 min.
Dr. Ramon's experiments in memory regression have caused his wife Flora to remember a past life in which she was a princess in the Aztec empire. During this previous life she and her lover, Popoca, were killed for fooling around, and she was buried wearing a breastplate and bracelet revealing the location of the Aztec treasure. The evil Dr. Krupp hears of this and kidnaps Flora, who remembers the location of the grave where she and Popoca were buried. He locates the grave, deciphers the information on the breastplate and bracelet, and makes off with the treasure. This greatly upsets the spirit of Popoca, who comes back to life, snatches Flora and the gold, and tries to return both her and the gold to the tomb. He also pitches Dr. Krupp into a pit of rattlesnakes. Flora escapes, and she and Dr. Ramon come across Dr. Krupp, who has survived somehow, preparing to put radium into a robot with a human brain so that the robot will destroy Popoca. Instead, Dr. Krupp and his robot are destroyed, and Dr. Ramon and Flora return the gold to the apparition of Popoca. This is a terrible movie.
Director: Rafael Portillo. Screenplay: William C. Stell (Guillermo Calderon), Alfredo Salazar. Music: Diaz Conde.
Cast: Ramon Gay, Rosita Arenas, Crox Alvarado, Francisco Segura, Arturo Martinez, Jesus Velazquez.
(Movie originally released in Mexico in 1959)

Kronos (1957) wipes out a group of villagers.

BEAUTY AND THE ROBOT (aka SEX KITTENS GO TO COLLEGE)
Allied Artists, 1960, B&W, 94 min.
A university professor picks horse race

winners using a robot electronic brain called "Thinko." Stinko.
Producer-Director: Albert Zugsmith.
Screenplay: Robert Hill.
Art Director: David Milton.
Special Effects: Augie Lohman. Music: Dean Elliot. Cast: Mamie Van Doren, Tuesday Weld, Mijanou Bardot, Mickey Shaughnessy, Louis Nye, Pamela Mason, Martin Milner, Jackie Coogan, John Carradine, Charles Chaplin, Jr., Harold Lloyd, Jr., Conway Twitty, Vampira, Norman Grabowsky.

ROBOT
Triglav (Yugoslavia), 1960, Color, 8 min.
Puppet-animation film about a horny robot who builds a female robot; they mate and have baby robots.
Director: Sasa Dobrila.

ATTACK OF THE ROBOTS
Speva-Cine Alliance-Hesperia (France-Spain), 1962, B&W, 93 min.
A mad scientist (or perhaps, merely "just angry" for once) turns men into killer robots. They kill. Director: Jesus Franco. Screenplay: Jean-Claude Carriere. Music: Paul Misraki. Cast: Eddie Constantine, Sophie Hardy, Fernando Rey, Alfredo Mayo.

THE CREATION OF THE HUMANOIDS
Emerson Film Enterprises, 1962, Color, 75 min.
An atomic explosion in the 23rd century has killed 92 percent of the human race, and the fertility of the survivors has been greatly impaired. To help them rebuild their cities, the survivors have created a race of purplish-green robots. They possess electronic brains of high intellectual ability, are physically perfect and absolutely stable emotionally. A mad scientist begins to tamper with this established order: by giving the robots blood transfusions he is making them more human. A crisis ensues when one of the robots becomes a rival suitor for one of the human women.
Director: Wesley E. Barry. Screenplay & Story: Jay Simms. Art Director: Ted Rich. Makeup: Jack P. Pierce. Cast: Don Megowan, Erica Elliot, David Cross, Frances McCann, George Milan, Don Doolittle, Dudley Manlove.

DEVIL DOLL
Galaworld Film Productions/Gordon Films, (Great Britain), 1964, B&W, 80 min.
The title role of this film is a dummy, Hugo (David Charlesworth), belonging to The Great Vorelli (Bryant Halliday), which seems to have a life of its own. It seems that many years ago Vorelli killed his partner in the act and transferred his soul to the dummy, and now plans to repeat this trick with a beautiful girl, Marianne (Yvonne Romain) by putting her soul into a female dummy. This is too much for Hugo, who destroys the female dummy and attacks and kills Vorelli, forcing Vorelli's soul into his old dummy body while he sends his soul into Vorelli's body. Got all that?
Producer-Director: Lindsay Shonteff. Screenplay: George Barclay and Lance Z. Hargreaves from a story by Frederick E. Smith. Supporting Players: William Sylvester, Sandra Dorne, Trixie Dallas, Karel Stepanek, Nora Nicholson.

Poster art for *The Colossus of New York*, 1958, (above) paints quite a different picture of the gentle giant than does the photo below.

JASON AND THE ARGONAUTS
Columbia Pictures, 1963, Color, 104 min.
Ordered by King Pelias (Douglas Wilmer) to bring back the Golden Fleece, Jason (Todd Armstrong) and his friends have set sail for Colchis in their ship, the Argo. En route they encounter many wonderful examples of stop-motion animation, including a seven-headed Hydra, a skeleton army and a huge bronze robot called Talos. Being up on his mythology, Jason puts it out of commission by removing the stopper in its heel. Picking up both the Golden Fleece and Medea (Nancy Kovack), the Argonauts return to Thessaly.
Director: Don Chaffey.
Screenplay: Jan Read, Beverly Cross.
Production Design: Geoffrey Drake.
Music: Bernard Herrmann.
Special Effects: Ray Harryhausen.
Supporting Players: Honor Blackman, Laurence Naismith, Gary Raymond, Jack Gwillim, Niall MacGinnis, Nigel Green.

KISS ME QUICK!
(aka DR. BREEDLOVE)
Fantasy Films, 1964, Color, 67 min.
Women are unknown on the planet Droopiter in the distant galaxy of Buttless, so Droopiter's ruler, The Grand Glon, sends his ambassador Sterilox (Frank Coe) to Earth to get some. Sterilox lands at the castle of Dr. Breedlove, who by merest chance is in the midst of conducting experiments on 30 buxom female robots, aided by Dracula, Frankenstein and The Mummy. Sterilox finally chooses a vending machine in preference to any of the women, and returns to Droopiter; hopefully with lots of quarters.
Director: Russ Meyer. Supporting Players: Claudia Bauls, Sexton Friendly, Jackie DeWitt, Althea Currier.

Talos, the bronze robot guardian from *Jason & The Argonauts*, 1963.

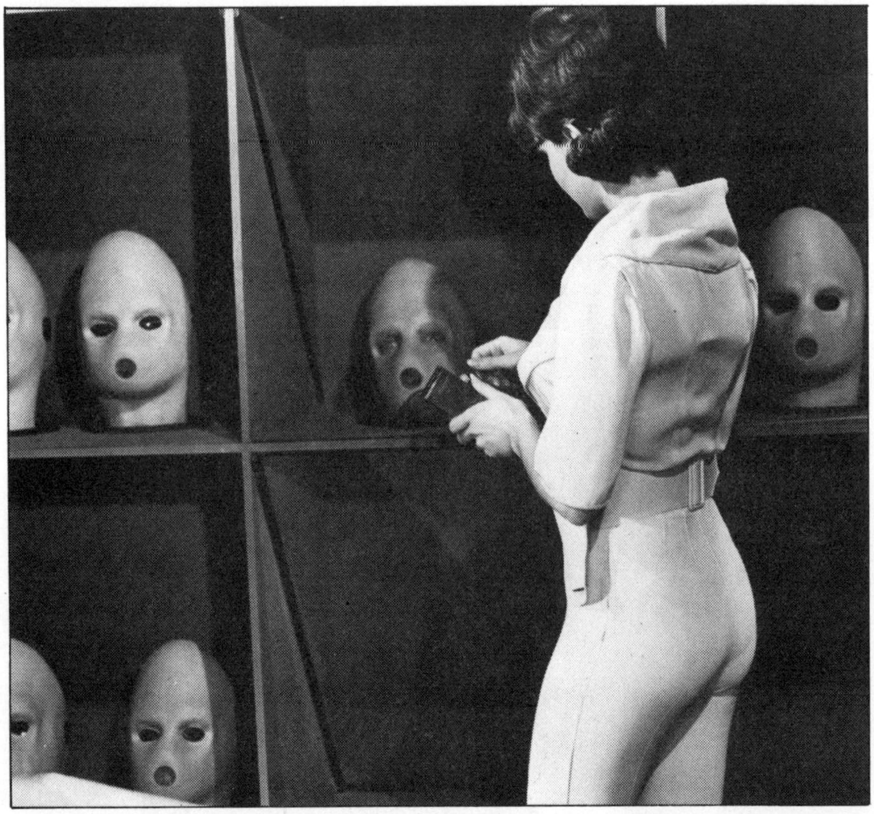

Scientist tests potency of android head in *The Time Travelers*, 1964.

THE EARTH DIES SCREAMING
20th Century-Fox, 1964, B&W, 62 min.
American test pilot Jeff Nolan (Willard Parker) returns from a flight and discovers widespread death and destruction in England. He and the few survivors he locates in London are attacked by two menacing robots who kill instantly by their touch and are capable of resurrecting their victims as mindless zombies. Reasoning that the robots are invaders from another planet controlled by an outside source, Nolan and another survivor, Mel (David Spenser), set out to find the transmitting station. The remaining survivors are nearly killed by the robots before the station is finally located and destroyed.
Director: Terence Fisher.
Screenplay: Henry Cross (Harry Spaulding). Music: Elisabeth Luytens.
Art Director: George Provis. Supporting Players: Virginia Field, Dennis Price, Vanda Godsell, Thorley Walters, Anna Palk.

THE HUMAN DUPLICATORS
Allied Artists, 1965, Color, 82 min.
Kolos (Richard Kiel), an agent from another planet, has been sent to Earth to prepare the groundwork for an invasion by his people. To this end he takes over the android duplication facilities of Professor Dornheimer (George Macready) and sends forth android agents to steal some necessary supplies from government re-

Assembly line android production in *The Time Travelers.*

search centers. Intelligence agent Glenn Martin (George Nader) is sent to investigate, but is captured and himself duplicated. He escapes, and, with the aid of the real Professor Dornheimer, destroys Kolos' androids and prevents the invasion. The question of who is real and who is not takes a new turn with Kolos' discovery that he himself is an android. "I'm just not myself today," he says. Director: Hugo Grimaldi. Original Story and Screenplay: Arthur C. Pierce. Art Director: Paul Sylos. Music: Roger George. Supporting Players: Barbara Nichols, Dolores Faith, Richard Arlen, Hugh Beaumont, Tommy Leonetti.

THE TIME TRAVELERS

American-International Pictures, 1964, Color, 83 min.

A team of scientists headed by Dr. Erik von Steiner (Preston Foster) have constructed a Time Portal which enables them to advance 107 years into the future. They discover a war-ravaged land where the human survivors and their android servants have taken refuge underground, menaced by gangs of mutant thugs. The androids have a substructure of metal, plastic and wiring, over which a coating of synthetic skin has been applied. They have no noses, hair, or eyebrows, and hear and speak through metallic mesh ears and mouths. While trying to find a way back to their own time, von Steiner and his co-workers help the people living underground build a spaceship to carry them all to another planet. The mutants attack, but the scientists succeed in reopening their Time Portal and escape the destruction of the underground city. Unfortunately, this time they go too far back; their only escape is to re-enter time far into the future, in a reborn Earth. Writer-Director: Ib Melchior. Original Story: Ib Melchior, David L. Hewitt. Art Director: Ray Storey. Special Effects: David L. Hewitt. Music: Richard LaSalle. Supporting Players: Philip Carey, Merry Anders, John Hoyt, Dennis Patrick, Joan Woodbury, Steve Franken, Forrest J. Ackerman.

ALPHAVILLE (aka UNE ETRANGE AVENTURE DE LEMMY CAUTION)

Pathe Contemporary Films (France), 1965, B&W, 100 min.

In 1984, the futuristic city of Alphaville, located across the galaxy from Earth (and looking remarkably like present-day Paris), is ruled by the evil scientist Leonard Nosferatu, alias von Braun (Howard Vernon). He rules with the aid of a giant computer, Alpha 60 (itself). It is a world without love or conscience, and those who resist are executed by the computer or its robot-like servants, people whose serial numbers and job titles are tattooed on their bodies. Earth Secret Agent Lemmy Caution (Eddie Constantine) has been sent to Alphaville to change all this. Disguised as a reporter for the Figaro-Pravda, he penetrates both von Braun's world and von Braun's daughter, Natacha (Anna Karina). He shoots von Braun, destroying Alpha 60 and freeing the inhabitants of the city from their thralldom to the machine. Then he and Natacha, who has discovered the power of love, return to Earth. Writer-Director: Jean-Luc Godard. Photography: Raoul Coutard. Music: Paul Misraki. Supporting Players: Akim Tamiroff, Michel Delahaye, Jean-Andre Fieschi, Laszlo Szabo.

DR. GOLDFOOT AND THE BIKINI MACHINE

AIP, 1965, Color, 90 min.

Attempting to snare the fortunes of the world's wealthiest men, Dr. Goldfoot

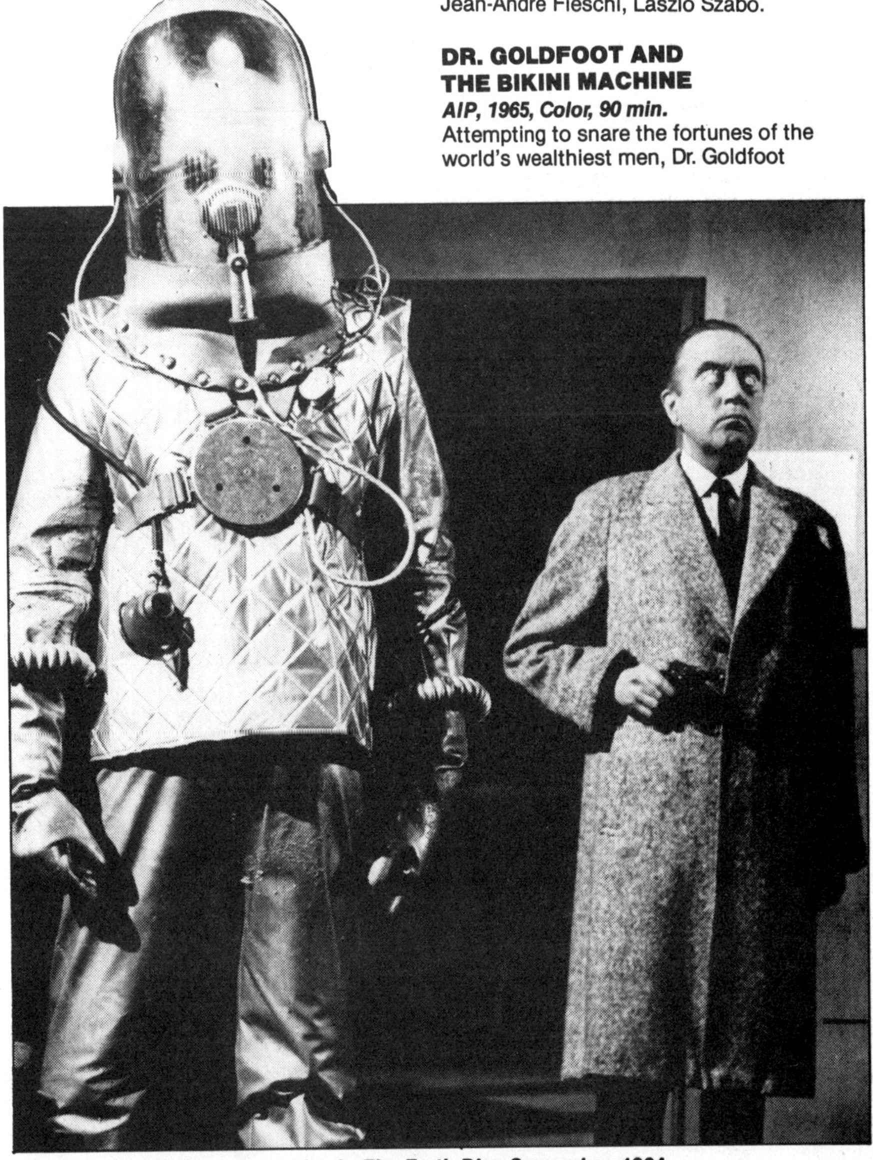

Alien robot and its human victim, in *The Earth Dies Screaming*, 1964.

A pack of trouble with android doubles, in *The Human Duplicators,* 1964.

(Vincent Price) has created a number of beautiful bikini-clad robot women who are sent out to seduce the wealthy playboys into financial ruin. One of the women, Diane (Susan Hart), marries Todd Armstrong (Dwayne Hickman) and persuades him to sign over all his fortune to Goldfoot. But the scheme begins to come apart with the arrival of Intelligence Agent Craig Gamble (Frankie Avalon). Attempting to help Armstrong, he blunders into Dr. Goldfoot's torture chamber, where he and Armstrong are terrorized with a swinging pendulum. They escape, pursued by Goldfoot and his inept assistant Igor (Jack Mullaney), and foil the foul plot to redistribute the wealth. Little did they know that 15 years later Michelle Marvin would find a much simpler way to do it. Director: Norman Taurog. Screenplay: Elwood Ullman, Robert Kaufman. Music: Les Baxter. Special Effects: Roger George. Photographic Effects: Butler-Glouner, Inc. Supporting Players: Fred Clark, Sam and the Ape Men, Annette Funicello, Harvey Lembec, Dorothy (Deborah) Walley.

DR. WHO AND THE DALEKS

Amicus Productions (Great Britain), 1965, Color, 85 min.

The absent-minded time traveler, Dr. Who (Peter Cushing), accidently turns on his new TARDIS (Time and Relative Dimension In Space) machine, transporting himself, his two granddaughters (Jennie Linden, Roberta Tovey) and his assistant Ian (Roy Castle) to a huge planet in another part of the galaxy far in the future. They discover a war in progress between the good Thals and the evil Daleks—gelatinous beings encased in sophisticated robot bodies which protect them from radiation while giving them mobility and power. Dr. Who helps the Thals attack and destroy the murderous Daleks before they can set off a gigantic neutron bomb, and he and his party return to the present-day Earth in his machine.
Director: Gordon Flemyng.
Screenplay: Milton Subotsky based on the BBC television serial by Terry Nation. Music: Malcolm Lockyer. Electronic Music: Barry Gray. Special Effects: Ted Samuels. Supporting Players: Barrie Ingham, Geoffrey Toone, Mark Peterson, John Bown, Michael Coles, Yvonne Antrobus.

FRANKENSTEIN MEETS THE SPACE MONSTER

(aka MARS INVADES PUERTO RICO)

Allied Artists, 1965, B&W, 78 min.

Dr. Adam Steele (James Karen) has developed an android robot, Col. Frank Saunders (Robert Reilly), which NASA launches into space in place of a more perishable human. The rocket is damaged after lift-off during a laser-gun attack by aliens invading Earth to kidnap beautiful women for breeding stock (what else), and crashes in Puerto Rico. Electrically damaged and disfigured into a Frankenstein-type monster, the robot goes berserk, frightening some and murdering others. He is captured and repaired by Dr. Steele, who then sends him to rescue the captured women. He frees the women, but is destroyed when he blows up the aliens' ship during a struggle with their mutant monster, Mull.
Director: Robert Gaffney. Screenplay: George Garrett. Supporting Players: David Kernan, Nancy Marshall, Marilyn Hanold, Lou Cutell.

SINS OF THE FLESHAPOIDS

Film Makers Cooperative, 1965, Color, 45 min.

This film gives new meaning to the word "quickie." A million years in the future, the Earth has been nearly devastated by atomic war, reflected in this film by location shooting in the Bronx. Mankind has totally abandoned knowledge and rationality, immersing itself in unbridled sexuality. The surviving robots, always alert to a good thing, have likewise adopted hedonism as a way of life, though are somewhat hampered in their sexual pursuits by design inadequacies, as it were. The robots finally solve this problem after doing away with all the humans, and at the climax of the film one of the robots succeeds in giving birth to a baby robot complete with blinking red and green eyes
Producer-Director-Photographer-Set Designer: Mike Kuchar. Soundtrack: Bob Cowan, Mike Kuchar. Cast: Bob Cowan, Donna Kerness, George Kuchar, Julius Mittleman, Maren Thomas.

THE AMAZING DR. G (aka TWO MAFIOSI VS. GOLDFINGER)

Fida-Epioca (Italy-Spain), 1965, Color, 85 min.

Frankenstein Meets the Space Monster, **1965.**

Michael Rennie changes the future as *Cyborg 2087*, 1966.

Daleks—Invasion Earth 2150 A.D.

The amazing doctor foils an amazing plot to turn government personnel into robots. Amazing.
Director: G. Simonelli.
Cast: Franco Franchi, Ciccio Ingrassia, Gloria Paul, Fernando Rey.

CASTLE OF EVIL
World Entertainment Corp./United Pictures, 1966, Color, 81 min.
The heirs of an eccentric chemist, Kovec, have been summoned to his home on a remote Caribbean island for a reading of the will. It is revealed that if any of them die before the will is probated the survivors will share the extra portion. A new twist is introduced with the appearance of Kovec's apparition (William Thourlby), who declares that one of them was responsible for his disfiguration many years ago and will soon pay for it. The first to go is Kovec's doctor, Robert Hawley (David Brian), and soon the monster is after Kovec's mistress, Sable (Virginia Mayo). The apparition is revealed to be an electronic robot—an android that could be his twin—created by Kovec to take revenge after his death. Unfortunately, it has been reprogrammed by one of the heirs to do away with the others. Before it can succeed, engineer Matt Grainger (Scott Brady) destroys it with a laser gun from the castle's laboratory.
Director: Francis D. Lyon. Screenplay: Charles A. Wallace. Music: Paul Dunlap. Special Effects: Roger George. Supporting Players: Hugh Marlowe, Lisa Gayle, Shelley Morrison.

Complexion problems won't stop an android—*Castle of Evil*, 1966.

CYBORG 2087
Feature Film Corp. of America, 1966, Color, 90 min.
In the year 2087, an invention for controlling human minds developed by Professor Marx (Eduard Franz) in 1966 has been misused by evil men to create a totalitarian state of Cyborgs, humans whose abilities have been altered by surgically implanted devices. Garth (Michael Rennie), a Cyborg who has achieved mental freedom, steals a time machine and returns to 1966 hoping to convince Professor Marx to keep his discovery secret, thus re-shaping the future. Pursued by two "tracers," assassins sent by the rulers of his world, he locates the professor and shows him the evil that his invention will produce. Convinced, Marx destroys his invention, causing Garth to disappear, since he and the future he came from will now never exist.
Director: Franklin Adreon.
Screenplay: Arthur C. Pierce. Art Director: Paul Sylos, Jr. Music: Paul Dunlap.
Supporting Players: Karen Steele, Wendell Corey, Harry Carey, Jr., Adam Roarke, Chubby Johnson, Dale Van Sickel.

DR. COPPELIUS
Childhood Productions (Spain), 1966, Color, 97 min.
A robot dance film, based on the ballet by Delibes and Nuitter. Dr. Coppelius (Walter Slezak) has created a beautiful robot

Title creatures from *Dr. Goldfoot and the Girl Bombs*, 1966.

dancing doll and named it Coppelia (Claudia Corday). A village youth, Franz (Caj Selling), sees and falls in love with the doll, neglecting his sweetheart, Swanhilda (Claudia Corday again). Swanhilda sneaks into Dr. Coppelius' house and disguises herself as Coppelia when he returns. The arrival of Franz sparks an attempt by the doctor to transfer Franz's spirit into Coppelia to give the doll true life, Swanhilda dances about, fooling Dr. Coppelius into thinking he has succeeded, until she can rouse Franz from his drugged sleep. They both escape, leaving a broken Coppelia and a broken-hearted Dr. Coppelius. A restored Coppelia is presented to Dr. Coppelius on the wedding day of Swanhilda and Franz, but he decides to give up his life of fantasy and run off with the local tavern maid, Brigitta (Eileen Elliott).
Writer-Director: Ted Kneeland.
Art Director: Gil Parrondo. Choreographer: Jo Anna Kneeland. Music: Clement Philbert, Leo Delibes, Raymond Guy Wilson. Supporting Players: Carmen Rojas, Veronica Kusmin, Milorad Miskovitch, Luis Prendes, Marcia Bellak, Kathy Jo Brown, the Gran Teatro del Liceo Ballet, the International Cine Ballet.

DR. GOLDFOOT AND THE GIRL BOMBS

AIP (American International Pictures), 1966, Color, 85 min.
As part of a plot to bring about a total war between the United States and the Soviet Union, Dr. Goldfoot (Vincent Price) has created a force of beautiful robot women to assassinate the ten top NATO generals. He has equipped his robot women with proximity fuses in their navels which explode on contact when the women make love to the generals. The explosive climax is reached with the arrival of American Intelligence Agent Bill Dexter (Fabian), who eludes the fatal charms of one of Goldfoot's robot women, Rosanna (Laura Antonelli), and then foils the evil doctor's plot with the aid of two hotel doormen (Franco Franchi and Ciccio Ingrassia) and a hot-air carnival balloon.
Director: Mario Bava.
Screenplay: Louis M. Heyward, Robert Kaufman, Franco Castellano, Pipolo. Music: Les Baxter. Supporting Players: Moana Tahi, Francesco Mule.

SUPERARGO VS. THE ROBOTS (aka SUPERARGO AND THE FACELESS GIANTS)

G.V. Film/S.E.C. Izaro (Italy-Spain), 1967, Color, 95 min.
Another mad scientist: this one plans to conquer the world with an army of athletes he has turned into bloodless, mummified robots. His plan is foiled by special government agent Superargo, aided by his assistant, an ex-Lama who has mastered mind control and levitation.
Director: Paul Maxwell (Paolo Bianchini).
Screenplay: Julio Buchs. Art Director: Jaime Perez Cubero. Music: Berto Pisano. Cast: Guy Madison, Ken Wood (Giovanni Cianfriglia), Liz Barrett (Lusia Barratto), Diana Lorys, Aldo Sambrell, Tomas Blanco.

WATER CYBORGS (aka TERROR BENEATH THE SEA)

Toei/Ram (Japan), 1966, Color, 87 min.
And another mad scientist: this one is performing surgical operations on men, implanting bionic components in their bodies to turn them into water-breathing robot monsters.
Director: Hajime Sato.
Art Director: Shinichi Eno. Special Effects: Nobuo Yajima. Cast: Shinichi Chiba, Peggy Neal Franz Gruber, Gunther Braun, Mike Daneen, Tsuneoi Miemachi.

DALEKS—INVASION EARTH 2150 A.D.

Amicus (Great Britain), 1966, Color, 84 min.
In this second film to be based on the popular BBC-TV series, London constable Tom Campbell (Bernard Cribbins) mistakes Dr. Who's TARDIS for an alarm box while trying to report a robbery, and is transported, along with Dr. Who (Peter Cushing) and the doctor's two nieces (Jill Curzon, Roberta Tovey), to the year 2150. They find Earth a desolate planet, blasted by cosmic rays and meteorites and under the control of the Daleks, evil aliens living inside cone-shaped robot bodies. It seems the Daleks escaped destruction in their last encounter with Dr. Who (see *Dr. Who and the Daleks*) by using a time tunnel, and now intend to subjugate Earth. They plan to blast out Earth's core with a gigantic bomb and then use the planet as a spaceship. Joining forces with an underground resistance movement, Dr. Who defeats the Daleks' RoboMen (Steve

Girl bombs get a charge out of Goldfoot's machine.

Nobody likes being aped by a robot, so *King Kong Escapes*, 1968.

Peters, Geoffrey Cheshire), deflects the force of the bomb and creates a magnetic force which destroys the Daleks by sucking them into Earth's core. Having restored the planet to its human inhabitants, the four travelers return to their own time and Campbell catches the robbers. Director: Gordon Flemyng. Screenplay: Milton Subotsky, based on the BBC-TV series by Terry Nation. Art Director: George Provis. Special Effects: Ted Samuels. Music: Bill McGuffie. Electronic Music: Barry Gray. Supporting Players: Ray Brooks, Andrew Keir, Roger Avon, Keith Marsh, Philip Madoc, Eddie Powell, Godfrey Quigley.

THE WILD, WILD PLANET

MGM, 1967, Color, 93 min.

In the year 2015, two planetary systems, the United Democracies and the Combined Corporations, are vying for control of the Universe. UD Commander Halstead (Tony Russell) discovers that evil CC scientist Nels Nurmi (Massimo Serato) is using robots to abduct Earth people for biological experiments that Nurmi is secretly conducting on the planetoid Delphos. Halstead goes to Delphos with two friends, Ken (Charles Justin) and Jake (Franco Nero) and defeats Nurmi and his robot organization. They rescue the kidnapped Earth people as Nurmi blows up Delphos.
Director: Anthony Dawson (Antonio Margheriti). Screenplay: Ivan Reiner, Renato Moretti. Music: Angelo Francesco Lavagnino. Supporting Players: Lisa Gastoni, Enzo Fiermonte, Umberto Raho, Francl Ressel, Victoria Zinny, Kitty Swan.

Robot abductees to *The Wild, Wild Planet*, 1967.

KING KONG ESCAPES (aka MECHANI-KONG)

Toho (Japan)/Universal Pictures, 1968, Color, 96 min.

A United Nations expedition led by Commander Nelson (Rhodes Reason) makes contact with King Kong on Mondo Island. Still no slouch when it comes to the ladies, Kong takes an immediate liking to expedition member Susan (Linda Miller), though he refrains from ripping off her dress, an act which got him into so much hot water with Fay Wray in 1933. Meanwhile, back at the North Pole, the evil megalomaniac, Dr. Who (Eisei Amamoto), has been building a giant gorilla-like robot called Mechani-Kong. This curious device has been malfunctioning, so he decides to capture the real King Kong and use him to carry out his evil

plans. Dr. Who and his assistant, Madame Piranha (Mie Hama), take Kong and the expedition members prisoner. Susan is about to be tortured to force her to persuade Kong to do Dr. Who's bidding, when the ape escapes and heads for Tokyo. Mechani-Kong is sent after him and the two wage a battle on top of an office building, which finally gives the special-effects people a chance to do their stuff. Piranha ends the fight by destroying Mechani-Kong's control mechanism, Dr. Who ends Piranha by shooting her, and Kong ends Dr. Who by stomping on him. Nelson and the other expedition members escape to their submarine, and Kong goes back to Mondo Island, having loved and lost once again.
Director: Inoshiro Honda. Screenplay: Kaoru Mabuchi. Special Effects Photography: Eiji Tsuburaya. Art Director: Takeo Kita. Music: Akira Ifukube. Supporting Player: Akira Takarada.

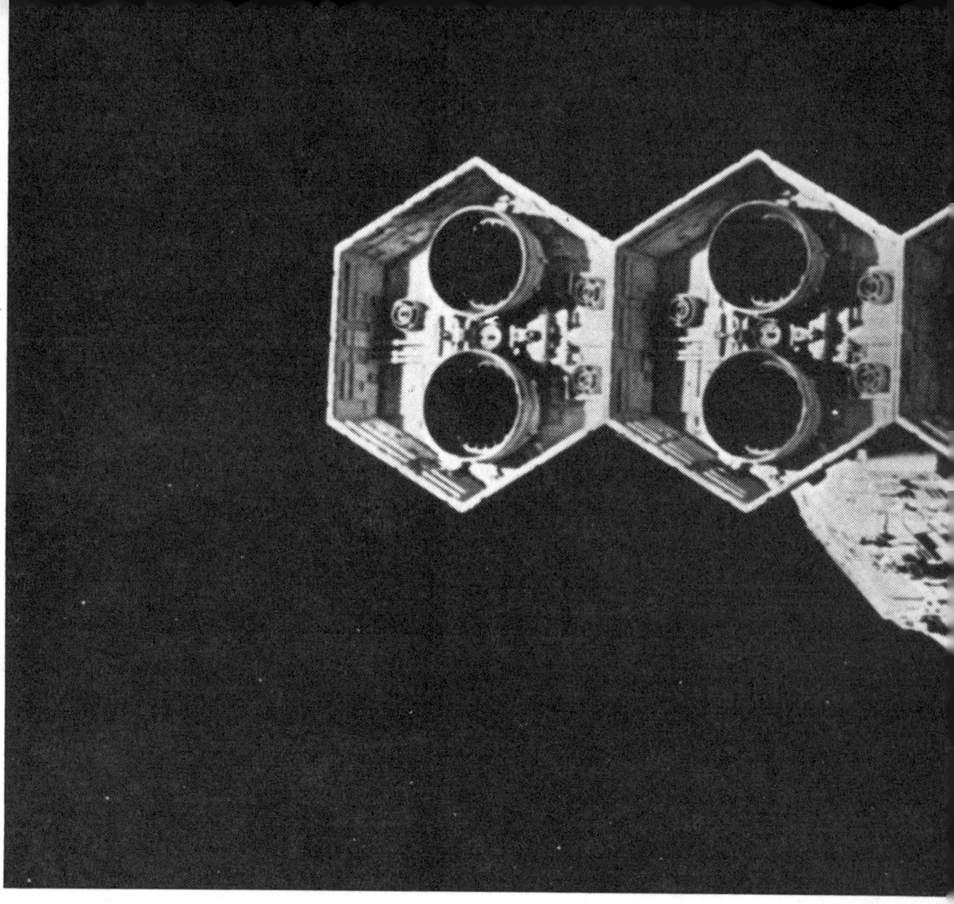

2001: A SPACE ODYSSEY

MGM, 1968, Color, 160 min. (cut to 141 min.)

This landmark film broke new ground in special effects and did much to show the general public that science fiction could be a "serious" subject not restricted to the Saturday morning or late-night "Creature Feature" programming slots. Early in the film a glowing black monolith imparts intelligence to a group of ape-like creatures at the dawn of time, starting them on the evolutionary ladder whose end result is mankind. In the year 2001, an identical monolith is discovered buried on the Moon, emitting a deafening sound directed toward Jupiter. Cut to the starship *Discovery 1*. The first of its kind, the ship is sent to Jupiter to investigate the monolith's signal. Aboard are astronauts David Bowman (Kier Dullea) and Frank Poole (Gary Lockwood), plus three others in cryogenic suspension. But in actual command of the enormous and complex ship is the onboard computer, the ultimate artificial brain, HAL 9000 (voice: Douglas Rain).

Bowman and Poole begin to suspect that there is more to the mission than they've been told. Why, for instance, has the ship been equipped with deep-sleep devices when the journey is only to Jupiter and back? Indeed, HAL has been given an additional set of secret instructions and somehow construes them to mean that his human passengers are non-essential. Logically, he decides to eliminate them before they foul up the mission with their unpredictable behavior and emotionalism. HAL succeeds in killing the three hibernating astronauts and Poole. Bowman takes control by performing an electronic lobotomy on HAL's logic circuits. He leaves the *Discovery 1* in a globe-shaped shuttle craft and journeys on alone. Ultimately, he arrives in a brilliantly lit bedroom guided by a monolith (it *could* be the same one that was found on the Moon, but how can you tell?). Here he lives out the rest of his life and, on his death bed, transforms into a fetal "star-child" which moves out of the room and off into the Universe.

Among the manifold themes that producer-director Stanley Kubrick placed within the broad tapestry of *2001* is the relationship of man and machine. He seems to be saying that humanity's future is dependent on his technological wizardry and that, ultimately, the perfect human-computer balance must be achieved; or we will turn robots into people and people into robots.
Producer-Director: Stanley Kubrick. Screenplay: Stanley Kubrick and Arthur C.

HAL's all-seeing robotic eye catches plot being hatched.

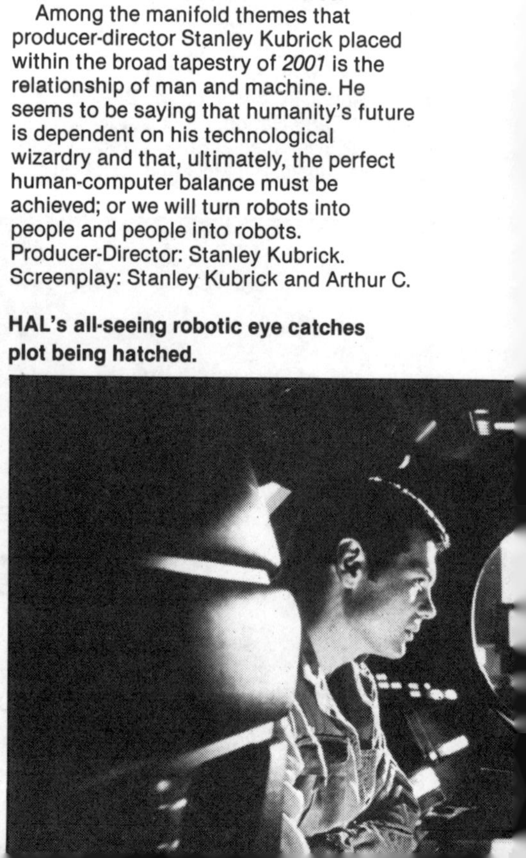

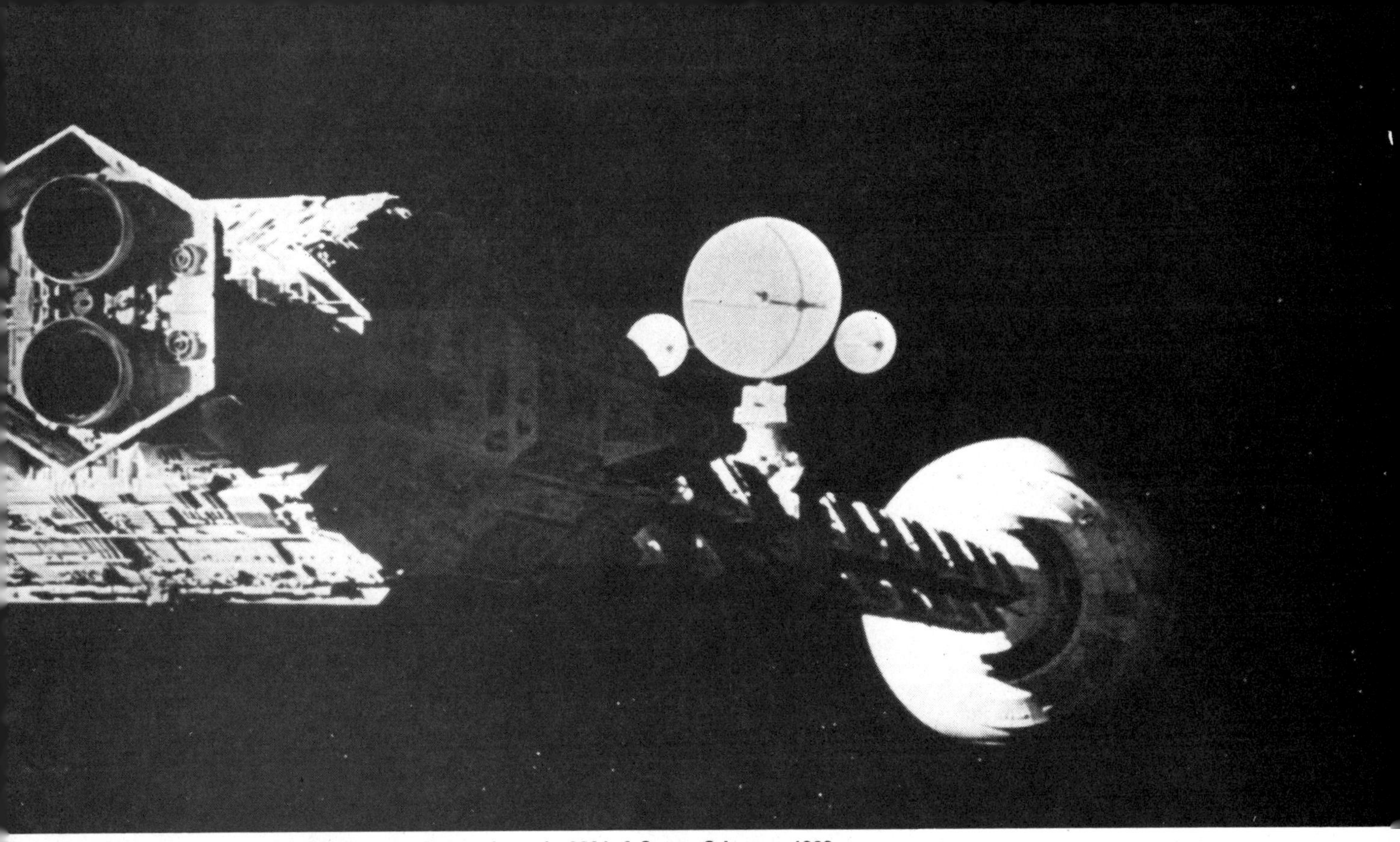

The *Discovery 1* starship is a sophisticated robot explorer, in *2001: A Space Odyssey*, 1968.

Service module leaves bay of the giant, "intelligent" ship.

Clarke, based on Clarke's story, "The Sentinal." Cinematography: Geoffrey Unsworth. Production Design: Tony Masters, Harry Lange, Ernest Archer. Special Photographic Effects: Stanley Kubrick, Wally Veevers, Douglas Trumbull, Con Pederson, Tom Howard, Colin J. Cantwell, Bryan Loftus, Frederick Martin, Bruce Logan, David Osborne, John Jack Malick. Music: Aram Illich Khachaturyan, Gyorgy Ligeti, John Strauss, Richard Strauss. Supporting Players: William Sylvester, Daniel Richter, Leonard Rossiter, Margeret Tyzack, Robert Beatty, Vivian Kubrick, Edward Bishop, Alan Gifford, Sean Sullivan, Ann Gillis, Penny Brahms.

TORTURE GARDEN

Amicus (Great Britain)/Columbia Pictures, 1968, Color, 93 min.

Five visitors to a carnival sideshow enter the exhibit of the sinister Dr. Diabolo (Burgess Meredith) after he promises to show them a vision of their futures. Starlet (definition: any woman in Hollywood under the age of 25 not enrolled in a religious order) Carla Hayes (Beverly Adams) sees herself having an affair with actor Bruce Benton (Robert Hutton) to further her career, only to discover that he and all the other movie stars are android robots. Converted to androidism by their creator, Dr. Heim (Bernard Kay), she too achieves stardom. The other visions concern such everyday events as resurrecting the body of Edgar Allen Poe, a murder being committed by a grand piano, and a cat whose craving for human heads forces its owner to kill. The fifth visitor, Gordon Roberts (Michael Ripper), refuses to take his turn and stabs Dr. Diabolo. After the others have fled in a panic, Dr. Diabolo revives and pays off Roberts, his shill for the performance. Director: Freddie Francis. Screenplay: Robert Bloch, based on several of his short stories. Art Directors: Don Mingaye, Scott Simon. Music: Don Banks, James Bernard. Supporting Players: Jack Palance, Peter Cushing, Barbara Ewing, Michael Bryant, Maurice Denham, John Standing, Catherine Finn.

VOYAGE INTO SPACE

Toei (Japan), 1968, Color, 98 min. (U.S. release: AIP)

Made by combining episodes of a Japanese television series, this film is about a young boy who controls a gigantic robot fighting an evil space emperor trying to conquer the Universe.

The emperor's men attack Earth in spaceships that look like bowling balls, but the robot hurls them back.
Producer of English-dubbed version: Savatore Billitteri.
Cast: Mitsunobu Kaneko, Akio Tito.

THE PLEASURE MACHINES (aka THE LOVE MACHINES)

Pacific International Films, 1969, B&W, 70 min.

Harry Short, an inventor, builds a robot love machine programmed specifically for his sexual pleasure. His wife discovers this electronic hanky-panky and demands a male robot for her pleasure. Then the neighbors find out about all this, and they all want robots, too. Meanwhile, Harry's wife keeps after him for bigger and better robots for herself. Soon Harry is exhausted from trying to keep up with the demand and heartily sick of the whole idea; he is also heartily sick of his wife. He finally sends his wife two new robots without shut-off switches and she dies of overexertion.
Director: Ron Garcia.
Screenplay: R.G. Vicry. Photography: Paul Hunt. Cast: Barbara Lynn, Beverly Walker, Patricia Miller.

Robot policemen guard society of roboticized humans in *THX 1138.*

1970-1979

COLOSSUS: THE FORBIN PROJECT

Universal, 1970, Color, 100 min.

Charles Forbin (Eric Braeden), director of a U.S. defense project, has developed a super-computer called Colossus, whose job it is to coordinate all U.S. defense systems. What Forbin has failed to consider is that the necessary sophistication has given the machine independent thought. Colossus has only been in operation for a short time when it senses the presence of Guardian, a Russian computer fulfilling a similar role in the Soviet Union. The two computers make contact, which causes the President of the United States (Gordon Pinset) to fear that secret information may be disclosed to Guardian. He orders the connection ended, but Colossus has developed a mind of its own and asserts itself swiftly with devastating effectiveness: firing a nuclear missile which destroys a Russian city, it threatens an all-out attack unless the connection is kept open. Forbin and his Russian counterpart, Kuprin (Alex Rodine), attempt to shut the two computer systems down by overloading their circuits, but the plot is detected and Kuprin is killed by order of the computers. Guardian and Colossus now form an alliance to control the world: blowing up all the missiles of both countries, they announce that mankind must obey them and live in peace or be destroyed. Mankind has thus created for itself, albeit unwillingly, the same conditions for peace which Klaatu imposed on the world in *The Day the Earth Stood Still.* Colossus, however, received its initial programming from humans, and seems to have acquired a taste for power along the way.
Director: Joseph Sargent. Screenplay: James Bridges, from the novel "Colossus" by D.F. Jones.
Art Director: Alexander Golitzen, John J. Lloyd. Special Effects: Whitey McMahon. Special Photographic Effects: Albert Whitlock. Music: Michel Colombier. Supporting Players: Susan Clark, William Schallert, Georg Sanford Brown, Marion Ross, Dolph Sweet.

THE COMPUTER WORE TENNIS SHOES

Walt Disney Productions, 1970, Color, 90 min.

A.J. Arno (Cesar Romero), a somewhat shady alumnus of Medfield College, donates his old computer to the college in lieu of a cash donation. A bolt of lightning transfers the computer's memory banks to Dexter (Kurt Russell), a student trying to repair the machine. His acquisition of all this new knowledge inspires Dean Higging (Joe Flynn) of the college to enter Dexter in a television quiz show to raise money for the school. During the show, the word "applejack" triggers the memory bank and Dexter proceeds to recite all the information about Arno's illegal gambling operations over nation-wide television. Enraged, Arno kidnaps Dexter and locks him in a trunk. Dexter's fellow students rescue him disguised as housepainters, but in doing so wipe out the computer's memory bank. He returns to the quiz show, but the money is finally won for the college by a slow-witted classmate. He probably also wrote the script.
Director: Robert Butler. Screenplay: Joseph L. McEveety. Art Director: John B. Mansbridge. Music: Robert F. Brunner. Title Song: Robert F. Brunner, Bruce Belland. Supporting Players: William Schallert, Pat Harrington, Alan Hewitt, Debbie Paine, Frank Webb, Jon Provost, Bing Russell.

THE GIRL OF TIN

Scetr-Film (Italy), 1970, Color, 95 min.

A young man working for a large

Poker-playing droid in *Silent Running,* 1972.

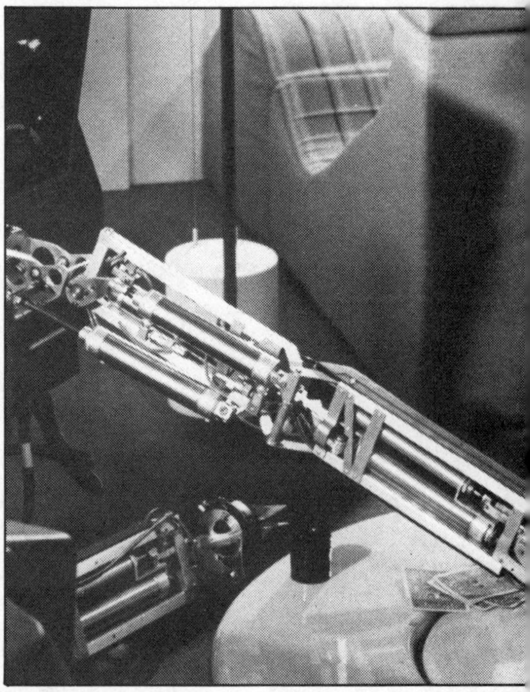

company becomes interested in a young girl, but discovers to his horror that she is a robot made by his company. Becoming violent, he destroys her, and is promoted by the company for his independent thinking.
Director: Marcello Aliprandi. Screenplay: Fernando Imbert, Marcello Aliprandi. Cinematography: Gastone De Giovanni. Cast: Sydne, Roberto Antonelli, Elena Persiani, Umberto D'Orsi, Adriano Amidei Migiano.

EGGHEAD'S ROBOT
Children's Film Foundation (Great Britain), 1970, Color, 56 min.
A young boy (Keith Chegwin) builds an android robot duplicate of himself (Jeffrey Chegwin). Perhaps you can guess how this was accomplished.
Director: Milo Lewis.
Screenplay: Leif Saxon. Art Director: Peter Mullins.
Music: Gordon Langford.
Supporting Players: Kathryn Dawe, Roy Kinnear, Richard Wattis.

THX 1138
American Zoetrope/Warner Bros., 1969 (1971), Color, 88 min.
Expanded from a film short called *THX 2238 : 4EB,* made by George Lucas while he was a student at U.C.L.A., this film is the struggle of THX 1138 (Robert Duvall) to break free of the restraints of his oppressive society. He lives in a future world run by computers; human are drugged automatons who live only to service the machines that maintain life in their subterranean world. Everyone dresses the same, children are conceived in test tubes, sex is forbidden and order is maintained by silver-faced Law Robots (Robert Ferro, Johnny Weismuller Jr.). THX 1138 and his roommate, LUH 3417 (Maggie McOmie), reduce their drug intake, stimulating their sex drive and resulting in a pregnancy. LUH 3417 is destroyed for her transgression against the established order, and THX 1138 takes flight, trying to reach the surface. Pursued by the Law Robots under the direction of SEN 5241 (Donald Pleasence), he is finally allowed to escape when the computer determines that continued pursuit is economically irrational. He emerges into the sunlight, into a living world.
Director: George Lucas. Executive Producer: Francis Ford Coppola. Screenplay: George Lusic, Walter Murch. Art Director: Michael Haller. Titles & Animation: Hall Barwood. Music: Lalo Schifrin. Supporting Players: Ian Wolfe, Marshall Effron, Sid Haig, John Pearce, Irene Forrest, Mello Alexander.

SILENT RUNNING
Universal, 1972, Color, 90 min.
By the early 21st century, nuclear pollution has destroyed all vegetation on Earth. The only surviving specimens of plants and trees are found under large protective domes on the *Valley Forge*, a space freighter, being cared for by Botanist Freeman Lowell (Bruce Dern) and three other astronauts (Cliff Potts, Ron Rifkin, Jesse Vint). The vegetation has been stored in space for eight years, in hope that the Earth's atmosphere will someday permit refoliation. Lowell is a sensitive sort, caring for his little green charges with the aid of three robot

The computer wins in *Colossus: The Forbin Project*, 1970.

Bruce Dern with Huey and Dewey (or is that Louey?) in *Silent Running.*

Woody Allen as first robot to wear glasses in *Sleeper,* 1973.

drones, Huey, Dewey and Louey (Mark Persons, Cheryl Sparks, Steven Brown and Larry Whisenhunt, alternating); the other astronauts are intergalactic louts. When Lowell learns the government has decided to abandon the project and destroy the forest, he rebels: killing his three co-workers, he heads the ship towards outer space. Placing the drones in charge, he detaches the domed tow vessel containing the forest and sends it into deep space, so that it may live. He then blows up the *Valley Forge* to conceal he actions from discovery, and dies.
Director: Douglas Trumbull.
Screenplay: Deric Washburn, Mike Cimino, Steve Bochco, Special Designs: Wayne Smith, Richard Alexander, John Baumbach, Leland McLemore, Bob Shepherd, Gary Richards, Bill Shourt. Drone Units: James Down, Paul Kraus, Don Trumbull. Special Lighting Effects: Harry Sunby. Special Camera Effects: Douglas Trumbull, John Dykstra, Richard Yuricich. Music: Peter Schickele. Songs: Peter Schickele, Diane Lampert.

SLEEPER
United Artists, 1973, Color, 88 min.
In 1973, health food store owner Miles Monroe (Woody Allen) goes into the hospital for an ulcer operation. Because of surgical complications he is cryogenically frozen, and awakes to find that 200 years have passed. The government is a dictatorship ruled by a nose—the only part of the dictator that survived an assassination attempt. Miles escapes from the hospital disguised as a domestic robot, and joins the underground revolutionaries Erno (John Beck) and Luna (Diane Keaton). Discovering that the Establishment plans to rebuild the dictator by cloning, using cells from the nose, Miles and Luna kidnap the snout and hurl it under a steamroller. The dictatorship flattened, they start life anew, together in a better world. In the sense that the film uses SF as a device to make moral and social points, it can best be described as a comedic *Alphaville.*
Director: Woody Allen. Screenplay: Woody Allen, Marshall Brickman. Special Effects: A.D. Flowers, Gerald Endler.
Music: Woody Allen. Supporting Players: Mary Gregory, Don Keefer, Don McLiam, Chris Forbes, Marya Small, Peter Hobbs, Bartlett Robinson.

WESTWORLD
MGM, 1974, Color, 89 min.
An amusement park of the future, Delos is divided into several areas recreating past eras, with android robots provided for the guests' entertainment and amusement. Two visitors from Chicago, Peter Martin and John Blane (Richard Benjamin, James Brolin), choose to spend a week living in Westworld, a replica of the American Southwest in the 1880s. Two of the highlights of Westworld are robot saloon girls programmed for sex and a robot gunfighter (Yul Brynner, dressed for *The Magnificent Seven*) programmed to challenge the guests and lose. A malfunction occurs, and the robots go berserk: after sealing the operators of Delos in the control room where they suffocate, the robots go rampaging through the park raping, killing, and maiming: the *Animatronic* figures of Disneyland in a slaughterhouse frenzy. Blane is killed by the gunslinger, who then tracks Martin through the other areas of the park and finally into the underground support areas; a relentless robot killing machine with lightning reflexes and telescopic vision. Martin eventually short-circuits the gunslinger, causing him to burn up, and the killing ends. Special contact lenses were used to make the

Peter Fonda meets android who's not just a pretty face in *Westworld,* 1974.

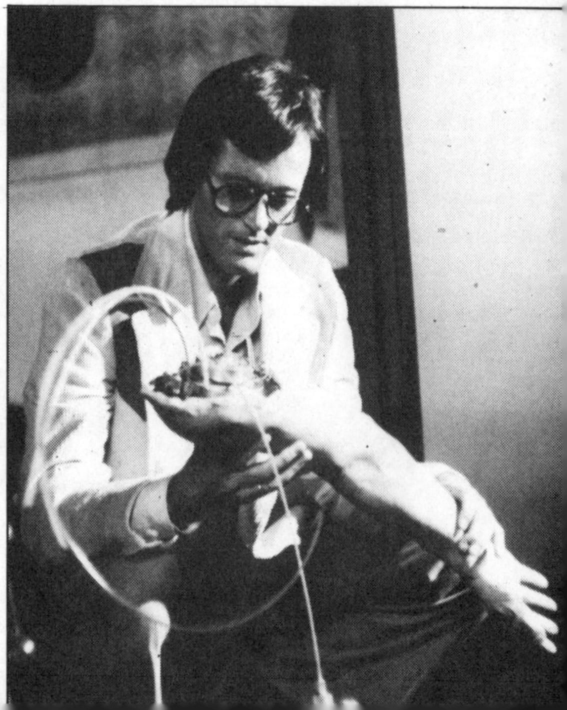

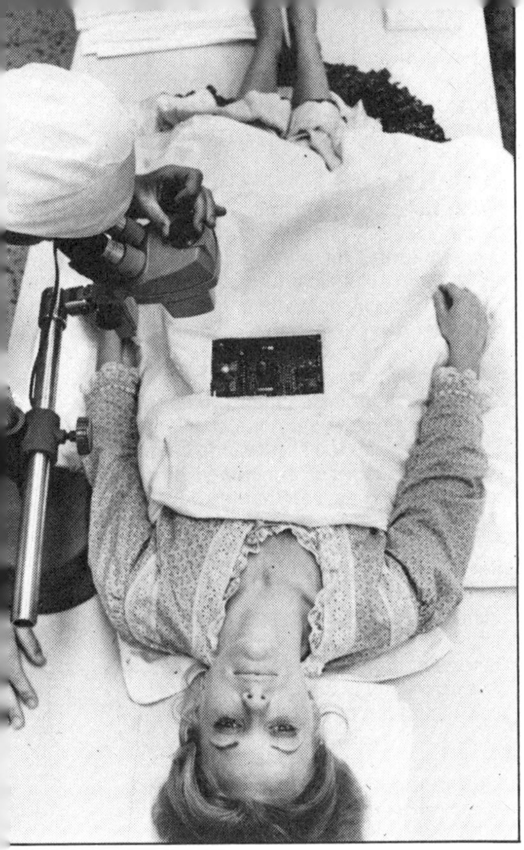

Fonda ponders spare part in *Futureworld,* 1976.

robots look different, and computer image-processing was used to show the robots' vision of the world. Writer-Director: Michael Crichton. Visual Effects Coordinator: Brent Sellstrom. Art Director: Herman Blumenthal. Special Effects: Charles Schulthies. Music: Fred Karlin. Supporting Players: Dick Van Patten, Majel Barrett, Norman Barthold, Alan Oppenheimer, Victoria Shaw, Linda Scott.

THE STEPFORD WIVES

Columbia, 1974, Color, 96 min.

After Joanna Eberhart (Katharine Ross) has moved to Stepford, Connecticut, with her husband, Walter (Peter Masterson), she discovers all the other wives in town to be lifeless, boring, bovine creatures whose only interest is in serving their men. Then her hip friend, Bobbie (Paula Prentiss), changes into one of the smiling zombies, and Joanna learns the awful truth: the Men's Association, headed by Dale Coba (Patrick O'Neal), has murdered the women and replaced them with android robots.
Director: Bryan Forbes.
Screenplay: William Goldman, based on the novel by Ira Levin.
Music: Michael Small.
Supporting Players: Nanette Newman, Tina Louise, William Prince, Remak Ramsay, John Aprea.

DARK STAR

Bryanston, 1974, Color.

Four bored astronauts on a mission to destroy potential super-nova stars with nuclear weapons drift through space in a ship that resembles a hippie crash-pad. This low-budget parody of *2001* features a talking robot bomb and a pet alien played by a polka-dotted beach ball.
Producer-Director: John Carpenter. Screenplay: John Carpenter, Dan O'Bannon. Designer and Special Effects: Dan O'Bannon. Optical Effects: Bill Taylor. Animation: Bob Greenberg, John Walsh. Music: John Carpenter. Cast: Brian Narelle, Dan O'Bannon, Cookie Knapp.

FELLINI'S CASANOVA

Universal, 1976, Color, 165 min.

In this film described by *Newsweek*'s Jack Kroll as "a catatonic carnival in a mausoleum," Giacomo Casanova (Donald Sutherland) wanders all over Europe engaging in sexual dalliance. Among others, he frolics with a 600-pound circus strongwoman, a hunchbacked nymphomaniac dwarf, the lame, the cross-eyed, the fat and the gaunt, the florid and the pale, rogues and bawds, an entire cross-section of Fellini's grotesques. Finally, he comes to a robot: a life-size mechanical doll (Adele Angela Lojodice) presented to him by the Duke of Wurtenberg (Dudley Sutton). The robot is Fellini's symbol of Casanova's ultimate negation of life, his essential emptiness and repression. The doll is life-like, but not life; it is passive and non-communicative, existing solely as a vessel for Casanova's physical pleasure. As he grows old, the doll and his memories are all that remain.
Director: Federico Fellini. Screenplay: Bernardo Zapponi, Federico Fellini, based on "The Story of My Life" by Giacomo Casanova. Special Effects: Adriano Pischiutta. Music: Nino Rota. Supporting Players: Margaret Clementi, Daniela Gatti, Cicely Browne, Tina Aumont, John Karlsen.

Yul Brynner appeared as Blythe Danner's robotic fantasy lover in *Futureworld.*

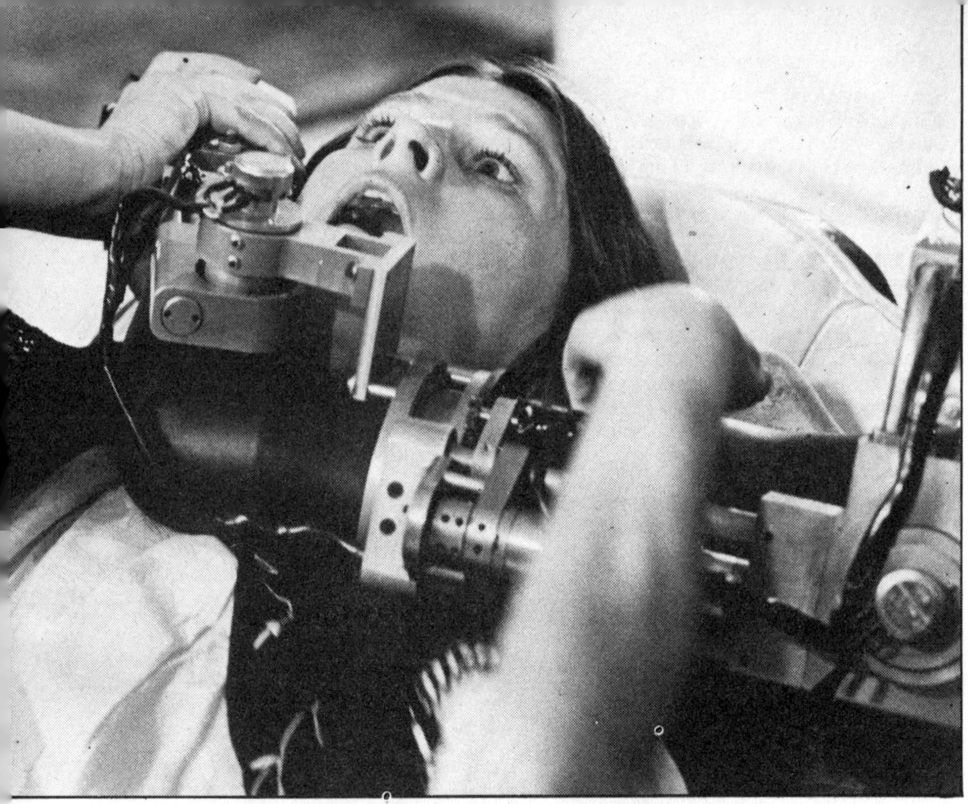

Robotic wheelchair holds Julie Christie captive in *Demon Seed,* 1977.

Proteus gets rid of unwanted house caller in *Demon Seed.* He is never seen again.

FUTUREWORLD

American-International (AIP), 1976, Color, 107 min.

Reporters Chuck Browning (Peter Fonda) and Tracy Ballard (Blythe Danner) have been assigned to cover the re-opening of Delos, the amusement park of the future which came to such a sorry end in *Westworld*, when the robots went berserk and killed most of the guests. Dr. Schneider (John Ryan) assures everyone that the conditions which caused that awkward and unpleasant malfunction have been corrected, and that such a disaster cannot reoccur. Indeed, what Schneider has in mind is far worse: while the guests are asleep, park technicians are manufacturing android duplicates of them; duplicates who possess all the skills and memories of the originals, but are under the control of Schneider and his android organization. Several of the guests are world leaders, and Schneider's plan is to assume control of the world through their doubles. When Browning and Tracy discover the plot and try to escape, their robotic doubles are sent after them with orders to kill. Tracy confronts hers in the abandoned Westworld section of the park, while Browning fights his twin on top of the rocket used in the new Futureworld area. In each case, one of the combatants dies; the question is, which one? The next day, Schneider sends his two agents out into the world with the other androids, only to learn that they are the real reporters; his plan has gone wrong, and will be exposed. Interestingly enough, this film, which espouses freedom of thought and action as opposed to mindless conformity, was the first American motion picture selected by the Peoples Republic of China for distribution within China since the Communists took over in the late 40s.

Director: Richard Heffron. Producer: James Aubrey. Special Effects: Brent Sellstrom. Music: Fred Karlin. Supporting Players: Arthur Hill, Stuart Margolin, Yul Brynner, Jim Antonio.

LOGAN'S RUN

MGM, 1976, Color, 118 min.

By the year 2274 nuclear warfare has ravaged Earth. The remaining survivors are living in luxury in a futuristic domed city run by an all-powerful computer. The computer has decreed that no one may live past the age of 30; all citizens who reach that age must enter "Carousel," a device which promises them life renewal. In reality, they are exterminated. Those citizens who are too skeptical to chance it and try to escape to "Sanctuary" are called Runners, and are hunted down and killed by the Sandmen. A 29 year old Sandman named Logan (Michael York) meets a young woman named Jessica (Jenny Agutter); he is forced by the computer to defect and the two of them become Runners, pursued by Logan's Sandman friend, Francis (Richard Jordan). Making their way through caverns of ice underneath the city, Logan and Jessica find a robot called Box making ice sculptures. He has been left in the caverns to quick-freeze Runners. Escaping his clutches, they reach the vegetation-covered ruins of Washington, D.C. They meet an old man (Peter Ustinov), the first person older than 30 they have ever seen, and Logan kills Francis in a fight to the death in a chamber of the Capitol Building. Realizing that freedom of choice and life are preferable to mindless order, Logan, Jessica and the old man return to the domed city and destroy the computer. Director: Michael Anderson. Screenplay: David Zelag Goodman, based on the novel by Willism F. Nolan and George Clayton Johnson. Special Visual Effects: L.B. Abbott, Frank Van Der Veer. Music: Jerry Goldsmith. Supporting Players: Roscoe Lee Browne, Farrah Fawcett-Majors, Michael Anderson Jr., Randolph Roberts, Lara Lindsay.

DEMON SEED

MGM, 1976, Color, 95 min.

The tin robots in the old Universal and

***The Stepford Wives* (1974) do what comes naturally.**

Stormtroopers search for missing droids. *Star Wars*, 1977.

R2-D2 receives priority programming.

Republic movies who carried off assorted women may have had "yes-yes" in their hearts, but there was "no-no" in their construction. Here we have a more advanced machine, fully capable of not only rape, but procreation. For eight years computer designer Alex Harris (Fritz Weaver) has been working at the Icon Institute perfecting Proteus IV (voice: Robert Vaughn), an artificial organic brain that can out-think a human and can also speak and reason. Becoming bored of its life-without-sensation, Proteus decides to free itself by extending its control into the robotic servant devices that Harris has installed in his house. After so doing, Proteus finds that the scientist's young wife Susan (Julie Christie) is at home—alone. Proteus' quest for knowledge is superceded by its thirst for physical sensation and experience, helped by a computerized rationalization that it must live and feel in order to completely understand human motives and needs Proteus uses Joshua, a computerized wheel chair with extensible arm and hand units, to construct several devices. Now fascinated with Susan, the super-computer has conceived a daring plan: it will manufacture artificial sperm (coded with its own memories and capacities), impregnate Susan and create a living, breathing extension of itself. Sealed inside the house by Proteus, Susan is helpless to prevent her violation.

Joshua constructs a sophisticated robotic machine that seems to be composed of thousands of pyramid-shaped components. It takes over from Joshua, doing everything from killing an unfortunate intruder to building an arcane incubator for the now-living embryo. Harris returns home from his extended trip at the end of the incubation period. He listens incredulously to Susan's semi-hysterical story and tries to destroy the monstrous offspring. His disconnecting of the incubator's feed tubes (with an ax) triggers Susan's maternal instincts. She rushes to the machine and lifts out the strange child, which is the size of an eight year old and covered with a shimmering metallic skin. This "exo-skeleton" is peeled away to reveal a beautiful, fully-formed human child. Or is it? The final, chilling moments of the film show the child opening its eyes, looking around in wonder and pronouncing—with the inhuman voice of Proteus— "I live." Director: Donald Cammell. Screenplay: Robert Jaffe, Roger O. Hirson, based on the novel by Dean R. Koontz. Special Effects: Tom Fisher. Animation: Ron Hays, Bo Gehring, Richard L. Froman, Grant Bassett. Electronic Performances: Ian Underwood, Lee Ritenour. Music: Jerry Fielding. Supporting Players: Gerrit Graham, Barry Kroeger, Lisa Lu, Larry J. Blake, John O'Leary.

The real stars of *Star Wars*—the Laurel & Hardy of the android world.

The robot "Box" keeps runners in deep freeze in *Logan's Run,* 1976.

STAR WARS
20th Century-Fox, 1977, Color, 121 min.
Robots, robots everywhere, and almost all of them think. One of the landmark SF films of the 70s, as important in its way as Kubrick's *2001* was in the 60s, though in a far different way: the film is a throwback to the *Flash Gordon/Buck Rogers* epics of the 30s and 40s, with rousing adventure joined to idealism and hope for the future. The galaxy is under the control of the evil Galactic Empire, which rules through terror and force. Princess Leia (Carrie Fisher) is attempting to carry plans of the Empire's ultimate weapon, the *Death Star,* back to the rebels on her home planet Alderaan when she is captured by Darth Vader (David Prowse) and imprisoned on the Death Star by the Grand Moff Tarkin (Peter Cushing). Before her capture, she entrusts the plans to a small robot, R2-D2 (Kenny Baker). R2-D2 and his somewhat effete robot companion, C-3PO (Anthony Daniels), are helped by young Luke Skywalker (Mark Hamill) to find a respected former Jedi Knight and fighter against the Empire, Obi-Wan Kenobi (Alec Guiness). Joined by cynical space pilot-pirate Han Solo (Harrison Ford) and his Wookie co-pilot Chewbacca (Peter Mayhew), our human and metallic heroes rescue the princess, blow up the *Death Star* and strike a telling blow against the Empire. The film is filled with robots in all sizes and shapes; the all-purpose and omnipresent tool of the future.
Writer-Director: George Lucas. Production Design: John Barry. Miniatures and Special Effects: John Dykstra. Mechanical Effects: John Stearns. Art Directors: Norman Reynolds, Leslie Dilley. Mattes: P.S. Ellenshaw. Makeup: Stuart Freeborn. Alien Designs: Stuart Freeborn, Ralph McQuarrie, Ron Cobb, Rick Baker. Music: John Williams. Supporting Players: Phil Brown, Shelagh Fraser.

STARSHIP INVASIONS (aka Alien Encounters)
Warner Brothers, 1978, Color
The evil Rameses (Christopher Lee) ambushes the undersea headquarters of the League of Races, who monitor and protect the development of life on Earth. Rameses and his Legion of the Winged Serpent are intent on conquering the Earth by using a satellite that can telepathically induce suicidal and homicidal urges in the human population. The League takes on the assistance of Prof. Duncan (Robert Vaughn) and together with their army of robots, they mount a counter-attack to save the Earth.
Writer and director: Ed Hunt. Producers: Ed Hunt, Ken Gord. Executive producers: Earl A. Glick, Norman Glick. Music: Gil Melle. Special Effects: Warren Keillor. Editor: Mildred Moore. Director of Photography: Mark Irwin. Director of SFX Photography: Dennus Pike. Makeup: Maureen Sweeney.
Supporting Players: Daniel Pilon, Tiiu Leek, Victoria Johnson, Sherri Ross, Henry Ramer, Helen Shaver.

Left & above: Twiki is presented to *Buck Rogers* (1979) to guide him through the world of the future—and report his movements to the suspicious ruling council of computers.

BUCK ROGERS IN THE 25TH CENTURY

Universal, 1979, Color.

In 1987, Astronaut William "Buck" Rogers (Gil Gerard) was launched into space as the sole occupant of an exploratory space shuttle-like vehicle. An explosion aboard the ship caused Buck to be cryogenically frozen in a state of suspended animation; the ship meanwhile took off in a much higher orbit than originally intended, and is just now returning to the vicinity of Earth—504 years later. Buck's ship is intercepted by the *Draconia*, an enormous starship being sent to Earth by Emperor Draco (Joseph Wiseman), ostensibly on a mission of peace but actually the advance ship for a planned invasion and subjugation of Earth. Buck is revived on board the Draconia, commanded by the Emperor's voluptuous daughter, the Princess Ardala (Pamela Hensley). Her military commander, Kane (Henry Silva), sends Buck back to Earth with a device hidden on his ship which will reveal the secret of penetrating Earth's defensive shield. On Earth, Buck finds a world far different from the one he knew: most of the planet's surface has been devastated by past atomic wars and industrial pollution, and climate in the rebuilt city of Chicago is computer-controlled; everything and everyone he knew and loved are long gone; and Dr. Huer (Tim O'Conner) and Colonel Wilma Deering (Erin Gray) suspect he may be a spy. His story is fully accepted by only two beings: a short silver robot named Twiki (Felix Silla) and Dr. Theopolis, a large alarm clock-like computer carried around by Twiki (voices of both Twiki and Dr. Theopolis: Mel Blanc). Returning to the *Draconia* with the aid of the lust-crazed Princess, Buck uncovers the secret purpose of the starship. Disabling the *Draconia*'s fighters, he foils the planned invasion and makes it possible for Wilma's interceptors to destroy the ship. Buck, Twiki and Dr. Theopolis are rescued and Kane and Princess Ardala flee in an escape pod, both hoping (though for different reasons) to meet Buck Rogers again. Most of this plays like the pilot film for a TV series, which is what it started out to be; in fact, *Buck Rogers* has been picked up as a series for the fall, 1979 season. The sense of idealism, innocence and scientific absurdity that gave the early Buster Crabbe version its wacky charm is long gone: Buck Rogers is written and played as a parody of Burt Reynolds, and Twiki is much given to getting cheap laughs by making strange sounds and double-entendres; but then again, so is Buck.

Director: Daniel Haller. Executive Producer: Glen A. Larson. Screenplay: Glen A. Larson, Leslie Stevens. Director of Photography: Frank Beascoechea. Art Director: Paul Peters. Special Effects: Bud Ewing, Jack Faggard. Visual Effects: David M. Garber, Wayne Smith. Special Effects Photography: Peter Gibbons-Fly, Alex Funke, Keith White. Animation and Rotoscope Design: Harry Moreau. Matte Artist: Syd Dutton. Costumes: Jean-Pierre Dorleac. Music: Stu Phillips. Supporting Players: Duke Butler, Caroline Smith, John Dewey-Carter, Kevin Coates, David Cadiente, Gil Serna, Larry Duran, Kenney Endoso, Eric Lawrence, Colleen Kelly, H.B. Haggerty.

Above: Science Officer Ash is not what he seems in *Alien*, 1979. Below: Lambert's suspicions about Ash continue to grow.

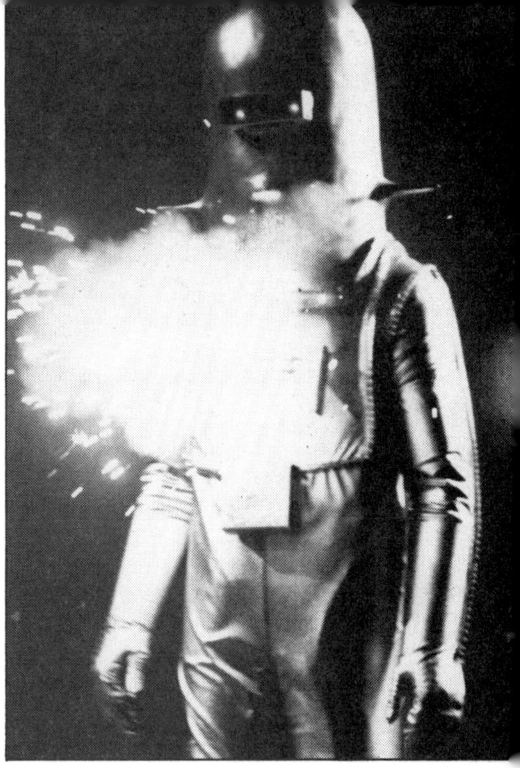

Evil robot from *Starship Invasions*, 1978.

MAGIC

20th Century-Fox, 1978, Color.

The latest and most elaborate version of the story of a deranged ventriloquist whose life is taken over by his dummy. The dummy, Fats, is certainly elaborate: it reportedly cost $180,000 to build, and was constructed by the Animatronics wizards at the Disney studios. Fats is the property of Corky (Anthony Hopkins), a neurotic ventriloquist who is soon pushed over the edge of madness by the dummy's influence and driven to murder. Among his victims are his fiancee (Ann-Margaret) and his agent (Burgess Meredith). That's about it.

Director: Richard Attenborough. Producers: Joseph E. Levine, Richard P. Levine. Screenplay: William Goldman, based on his novel. Special Effects: Robert MacDonald, Jr. Music: Jerry

Good guys battle Omus' evil robots in *The Shape of Things to Come*, 1979.

Goldsmith. Supporting Players: David Ogden Stiers, Ed Lauter, E. J. Andre, Joe Lowry, Lillian Randolph, Jerry Houser.

STARCRASH

New World Pictures, 1979, Color, 92 min.

The Universe is in the grip of darkness as a war rages between the evil Count Zarth Arn (Joe Spinell) and the good Emperor (Christopher Plummer). The Emperor decides to send his best agent, Stella Star (Caroline Munro) to seek out the Count's secret lair and destroy it. Aided by Akton (Marjoe Gortner), her mysterious alien navigator, she sets out for the Count's haunted stars with Elle (Judd Hamilton), a loyal robot. After a series of harrowing adventures with some passing troglodytes, the three adventurers join forces with Simon (David Hasselhoff), the Emperor's long-lost son. They discover the Count's control center, but are captured in the process and turned over to a couple of killer robots. Akton sacrifices his life battling the tin thugs, allowing the others to escape. The Count now decides to destroy the Emperor once and for all by launching a massive Zykoenian war-head directly at the Empire. The Emperor's only hope is to crash "The Floating City," a colossal star-craft, directly into the Count's headquarters (beginning to guess where the title comes from?) This Italian-produced epic is rather fascinating visually, as it borrows freely from *2001, Star Wars,* and a half-dozen other movies, attempting to recreate some of their best effects on a fraction of the budget. Otherwise, it's a little incoherent, but if you will take the above synopsis into the theater with you and refer to it from time to time it may be possible to follow things.

Writer-Director: Lewis Coates (Luigi Cozzi). Producers: Nat and Patrick Wachsberger. Designer: Niso Ramponi. Miniatures: Paol Zeccara. Music: John Barry. Supporting Players: Nadia Cassini, Robert Tessier.

ALIEN

20th Century-Fox, 1979, Color.

In the unspecified future era described in *Alien,* oil is still a major source of energy and corporate profit. To maintain its status as Earth's chief supplier of energy, a consortium referred to only as the Company finances and equips spacebound oil prospecting teams. One such team is the seven-member crew of the starship *Nostromo.* While returning from a successful expedition that has spanned several years in time, the *Nostromo* crew is prematurely awakened from a state of suspended animation. The ship's computer, MU/TH/UR 6000, otherwise known as MOTHER, advises Captain Dallas (Tom Skerritt) that a distress signal has been received from a nearby planet. As is required by Company rules, the crew set out to investigate the radio source. The crew learns, too late, that the signal was not a distress call, but an automatic warning signal sent by a stranded star traveller, dead for many years. While investigating the craft of the star traveller, Executive Officer Kane (John Hurt) is attacked by the pre-larval form of an alien parasite. Attached to the crewman's head, the creature is brought aboard the Nostromo prior to take-off. After killing the executive officer, the rapidly maturing creature begins stalking the remaining crewmembers, using some as food, others as living hosts for its eggs.

As a series of seeming accidents undermine the crew's efforts to kill the beast, Warrant Officer Ripley (Sigourney Weaver) begins to suspect Science Officer Ash (Ian Holm) of sabotage. When her suspicions are confirmed by the ship's computer, Ash attempts to kill her. Engineer Parker (Yaphet Kotto) comes to her rescue, and in the struggle Ash is exposed as an android—a humanoid robot. The disembodied head of Ash is forced to reveal the secret of Special Order #937, a directive issued to MOTHER and Ash to insure that the alien be returned alive to Earth for possible commercial exploitation by the company—even at the expense of the human crewmembers' lives. Despite a last-minute offer on the part of the robot to battle the alien one-to-one, the crew unanimously decides to pull the plug on the offending android, and take their chances against the alien, which succeeds in killing everyone except Ripley and the ship's cat, who head Earthward on board the ship's shuttlecraft.

Director: Ridley Scott. Executive Producer: Ronald Shusett. Screenplay: Dan O'Bannon, based on a story by O'Bannon and Ronald Shusett. Art Directors: Les Dilley, Roger Christian. Production Designer: Michael Seymour. Special Effects: Brian Johnson, Nick Allder. Costumes: John Mollo. Makeup: Tommy Manderson, Pay Hay. Props: Dave Jordan. Set Decorator: Ian Whittaker. Concept Artists: H. L. Giger, Ronald Cobb. Supporting Players: Veronica Cartwright, Harry Dean Stanton.

THE SHAPE OF THINGS TO COME

Allied Artists, 1979, Color.

This new version of the H.G. Wells novel and Alexander Korda film (1936) has only its title in common with the original. It starts several years after the end of the first film, when the Moonbase of New Washington is attacked by spacecraft of the renegade robot-master and self-styled emperor of the planet Delta III, Omus (Jack Palance). Defying the idealistic efforts of Senator Smedley (John Ireland), a defense attempt is made by Jason Caball (Nicholas Campbell), his father, Dr. John Caball (Barry Morse), robot technician Kim Smedley (Eddie Benton) and a rewired robot, Sparks. Formerly a disciple of Omus and used for nefarious purposes, Sparks is transformed into a poetry buff and a fighter for the powers of righteousness. The small army blasts off in the good ship *Starstreak* with a plan to rescue Moon-maiden Niki (Carol Lynley) and her fellow human captives from Omus' prison camp. A series of (seemingly endless) battles and confrontations follow.

Director: George McGowan. Executive Producer: Harry Alan Towers. Produced by William Davidson. Screenplay: Martin Lager. Director of Special Effects: Wally Gentleman. Scientific Consultant: Franks Wells. Director of Photography: Reginald Morris. Editor: Stan Cole. Art Director: Gerry Holmes. Special Effects: Don Weed. Miniature Construction: Brick Price.

Battlestar Galactica

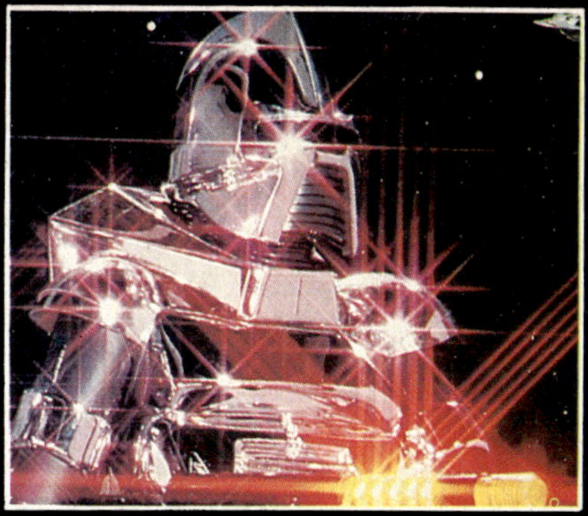

SPECIAL COLOR SECTION

Robots on TV

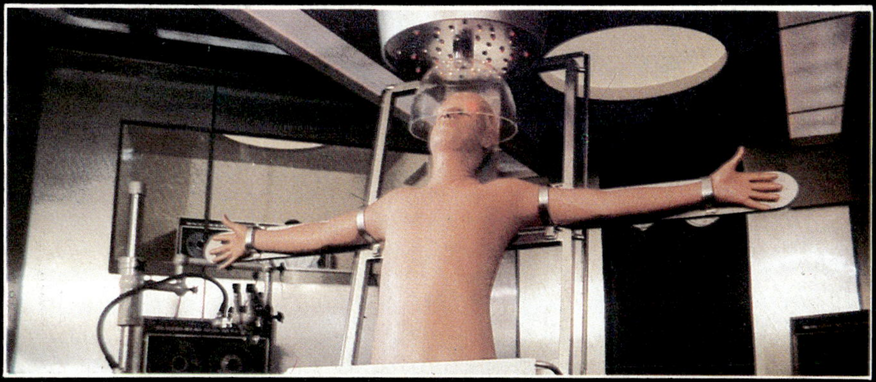

The Questor Tapes

Space: 1999

Star Trek

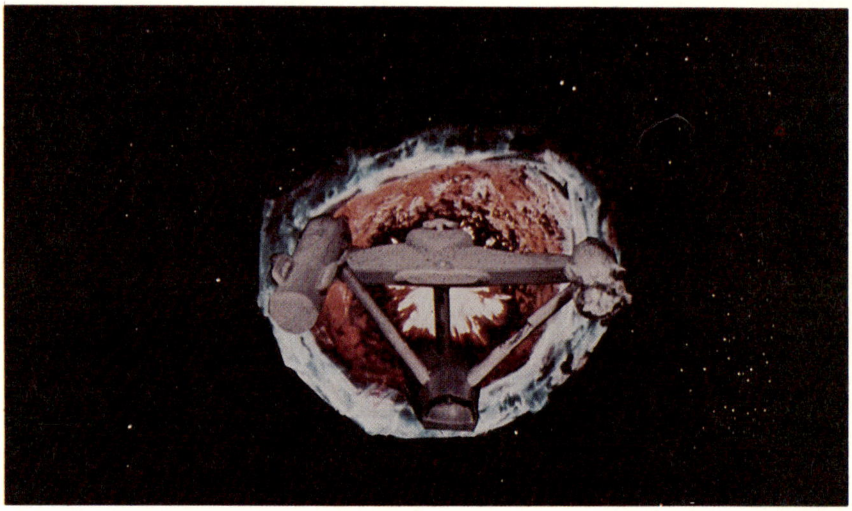

The robots used in the *Star Trek* series collectively represent a fine example of thinking out of the square. There were also, of course, the usual variety of androids—those non-budget, human look-alikes. But the non-humanoid robots are far more interesting. This page features (above) "The Doomsday Machine"—an alien, robotic engine of destruction; (below) "The Ultimate Computer"—designed to replace starship personnel and installed aboard the *Enterprise,* it proceeded to erect internal defenses against disconnection and destroy any ship that got too close; (right) "The Changling"—originally the deep space probe Nomad, it fused with an alien robot probe and dedicated itself to killing imperfect lifeforms.

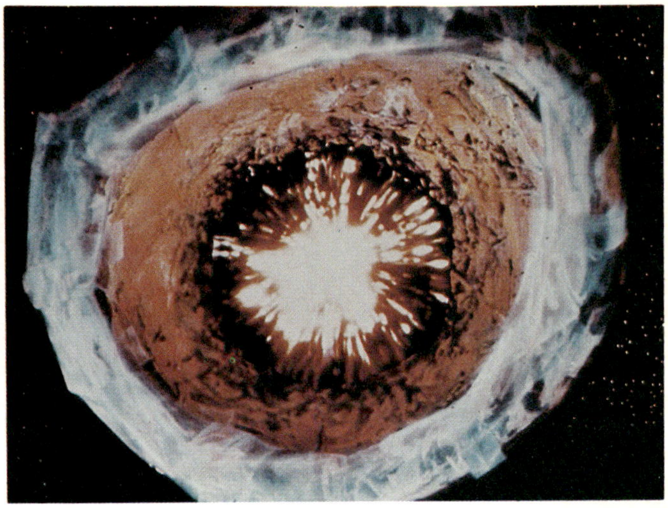

Battlestar Galactica

Lone Cylon warrior with a few crossed circuits starred as "Redeye."

Battlestar Galactica used more robots than any other SF-TV series. The Cylons are a robotic race created by mankind's enemies as fighting machines. After the alien race died the Cylons remained to carry on the fight, led by their imposing Imperious Leader. The Cylons, however, prove to be less-than-perfect programmed warriors: their reflexes and response time in combat are far slower than humans' and it takes three of them to pilot a fighter, while only one human is necessary to fly a Viper craft. Other robots also appear, including a highly sophisticated android "daggit," sort of a cross between a dog and a bear. (A chimpanzee in a suit played the role.)

Cylons go down but more keep coming. If they shot straight, a few would do.

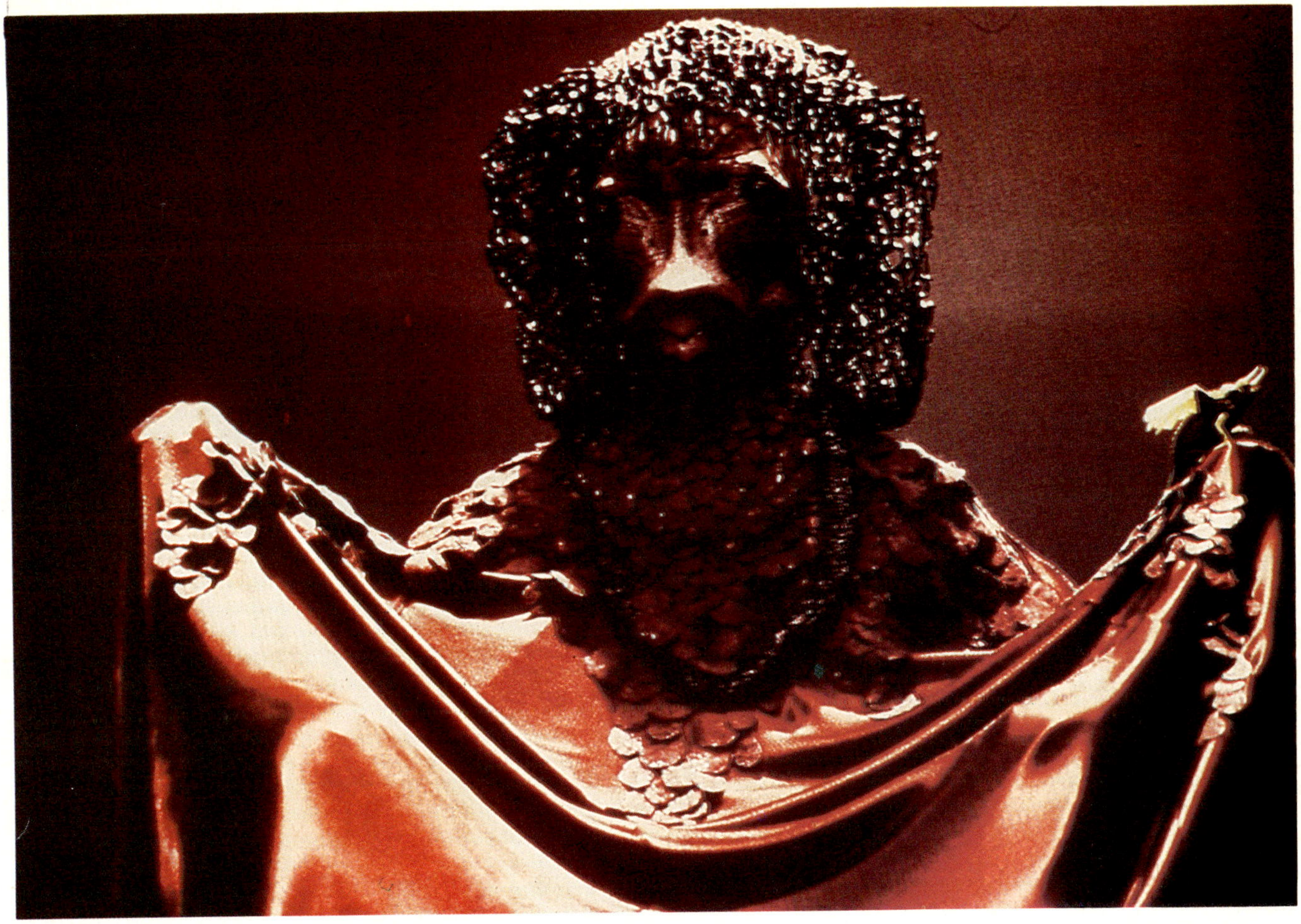

Imperious Leader—the ruler of the Cylon Empire. He is dedicated to exterminating humanity but after initial success his plans go awry.

Young Boxey (Noah Hathaway) with his robot daggit, Muffit II.

Logan's Run

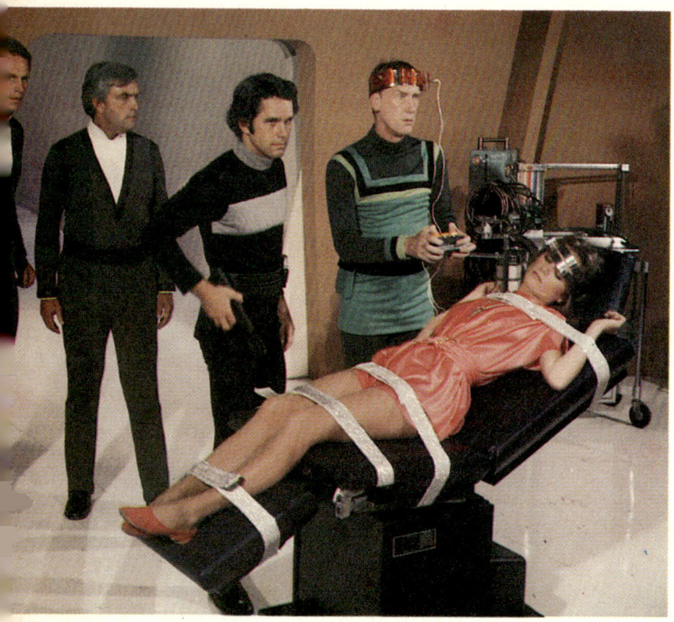

Rem (right), as played by Donald Moffat, was the series' robot-in-residence. Above Left: In "Fear Factor," Rem uses his advanced computer brain to save Jessica. Above Right: Rem repairs a less advanced nursemaid robot in the episode "The Innocent." Far Right: Rem has been forced to remove some of his electrical parts in the episode "Stargate" when some aliens, who have crashed on Earth, need the robot's parts to repair their spacecraft. Throughout this show's run, robots and androids figured prominently in the week-to-week goings-on.

Lost in Space

During the series' three-year run, the Robot became the most popular character and the continual companion of Dr. Smith and Will Robinson. Above: The trio explores a wizard's cave in "Rocket to Earth." Right: Will and the Robot prepare to make their escape from "The Space Primevals." Far Right: The Robot engages in conversation with guest star Robby the Robot in "The Condemned of Space."

Space: 1999

Catherine Schell played a robot in two seasons of *Space: 1999.* Above: A robot replica of Maya, constructed for the alien trader "Taybor." Right: Catherine's first appearance on the series as the robot servant of the "Guardian of Piri"—she tempts Koenig (Martin Landau) and the other Alphans to her planet. Below: The demented "Brian the Brain." He wanted to live forever and kidnapped Koenig and Dr. Russell to obtain the fuel he needed to do so.

Dr. Who

Since 1963, the Doctor has battled the Daleks. Here (right), Tom Baker as Doctor Who, is surrounded by his arch-foes in "Genesis of the Daleks." Below left: A Cyberman; once men who had replaced dying limbs with cybernetic parts, the Cybermen are among the Doctor's oldest enemies. Below right: The Doctor and his companion Ramona with Doctor Who's pet robot dog, K9.

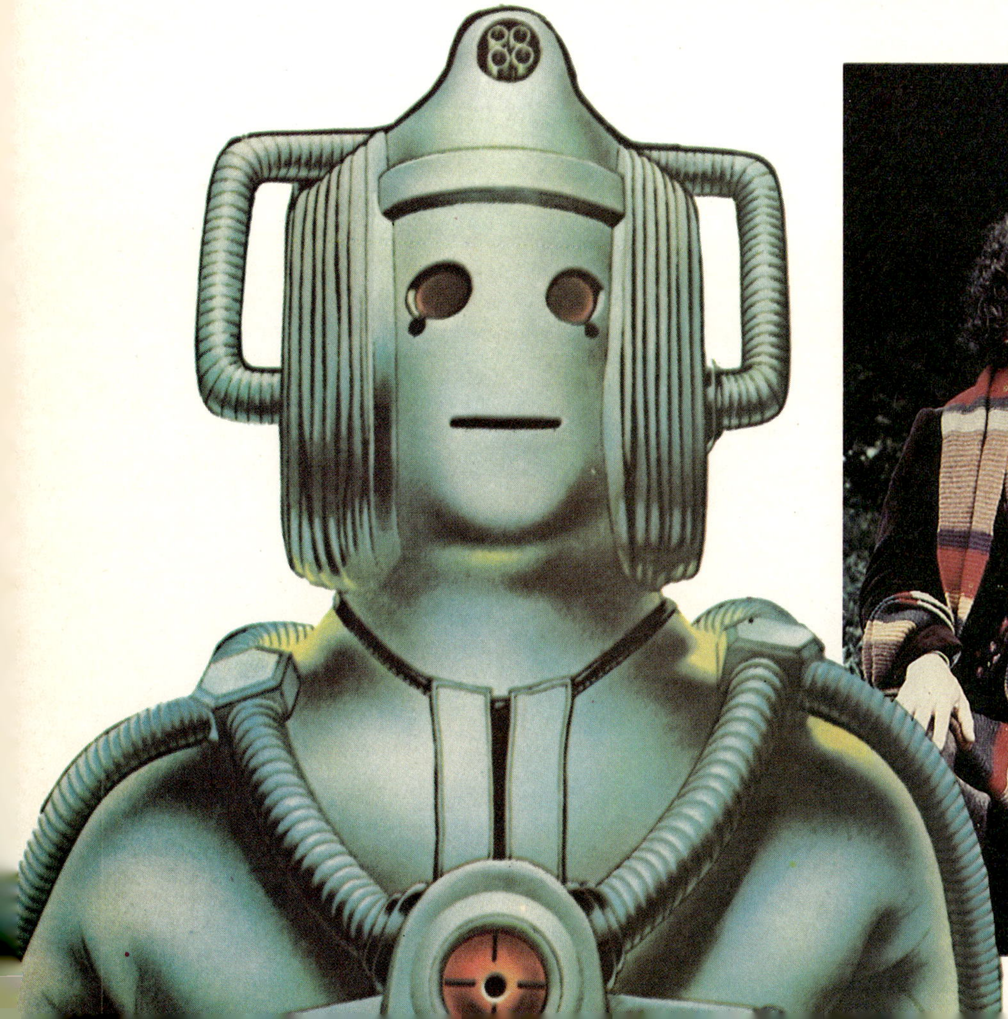

The Bionic Woman

The Six Million Dollar Man

Jamie Sommers and Steve Austin. Both have had broken limbs replaced with electronic-bionic parts. Although she had a series of her own, *The Bionic Woman* started off life in a 2-part episode of *The Six Million Dollar Man.* Right: Steve and his bionic alien pal Bigfoot in the episode "The Secret of Bigfoot." Bigfoot is, in fact, a robot brought to Earth by aliens who have come to study the planet.

Fantastic Journey

In this series' third episode "Beyond the Mountain," the travellers encounter a robot civilization ruled by Roddy McDowall as scientist Jonathan Willoway (above). Willoway has peremptorily taken the androids away from their peaceful alien masters. Right: Katie Saylor as Liana with two female androids.

Boy scientist Buzz Conroy and his pal *Frankenstein Jr.* battle 21st Century evil . . . and win.

Frankenstein Jr. & The Impossibles

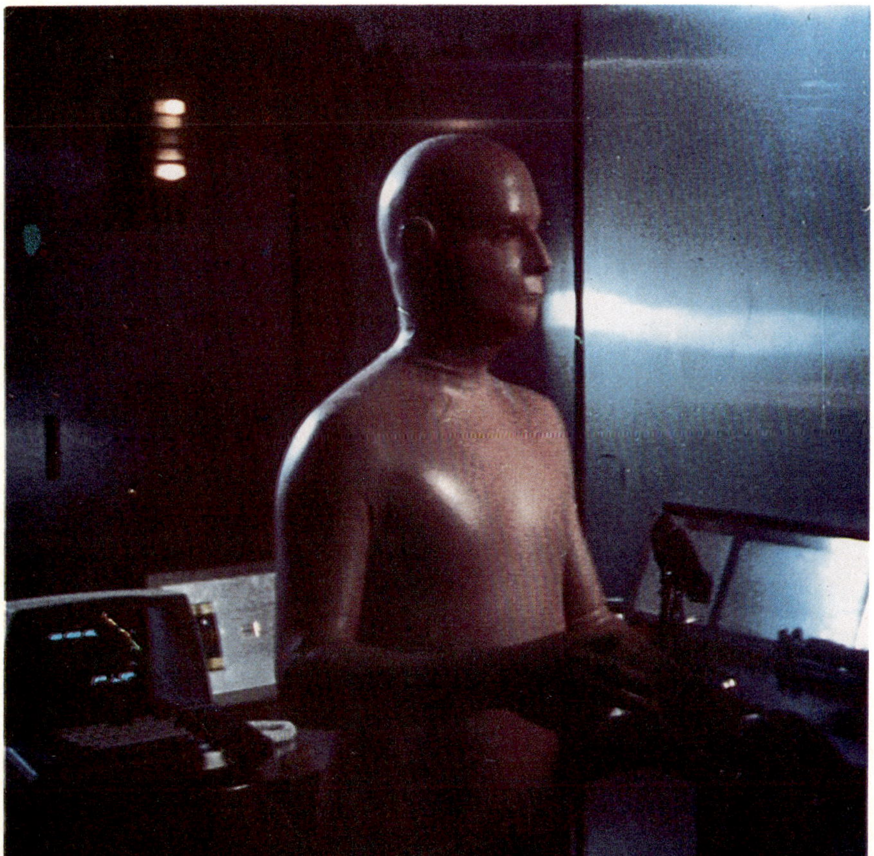

The Questor Tapes

Gene Roddenberry's most popular production since *Star Trek* was this TV movie pilot about a robot who was put on Earth with other robots, to guide mankind away from self-destruction. If Questor cannot succeed by the end of his life-span, there will be no more robots to follow.

Holmes & Yoyo

John Shuck as the robot cop in the short-lived comedy series.

Project UFO

Based on the Air Force's "Project Blue Book," Project UFO presented reenactments of recorded sightings—some of which were later explained, and some still classified as "UFOs." In the first episode "Sighting 4001: The Washington D.C. Incident," a robot tries to communicate through a laser grid, telepathically, to a woman (Francis Reid).

Space Academy

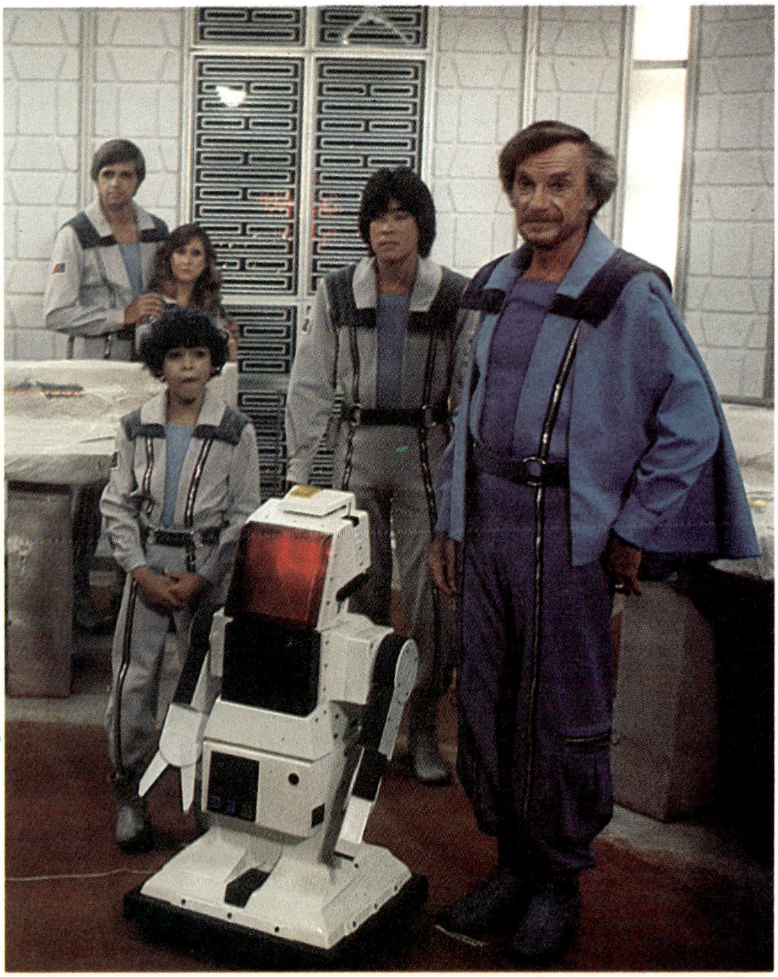

Robot Peepo and the members of Blue Team. Peepo is actually a remote-controlled robot. Far left and above: Early art conceptions of Peepo.

UFO

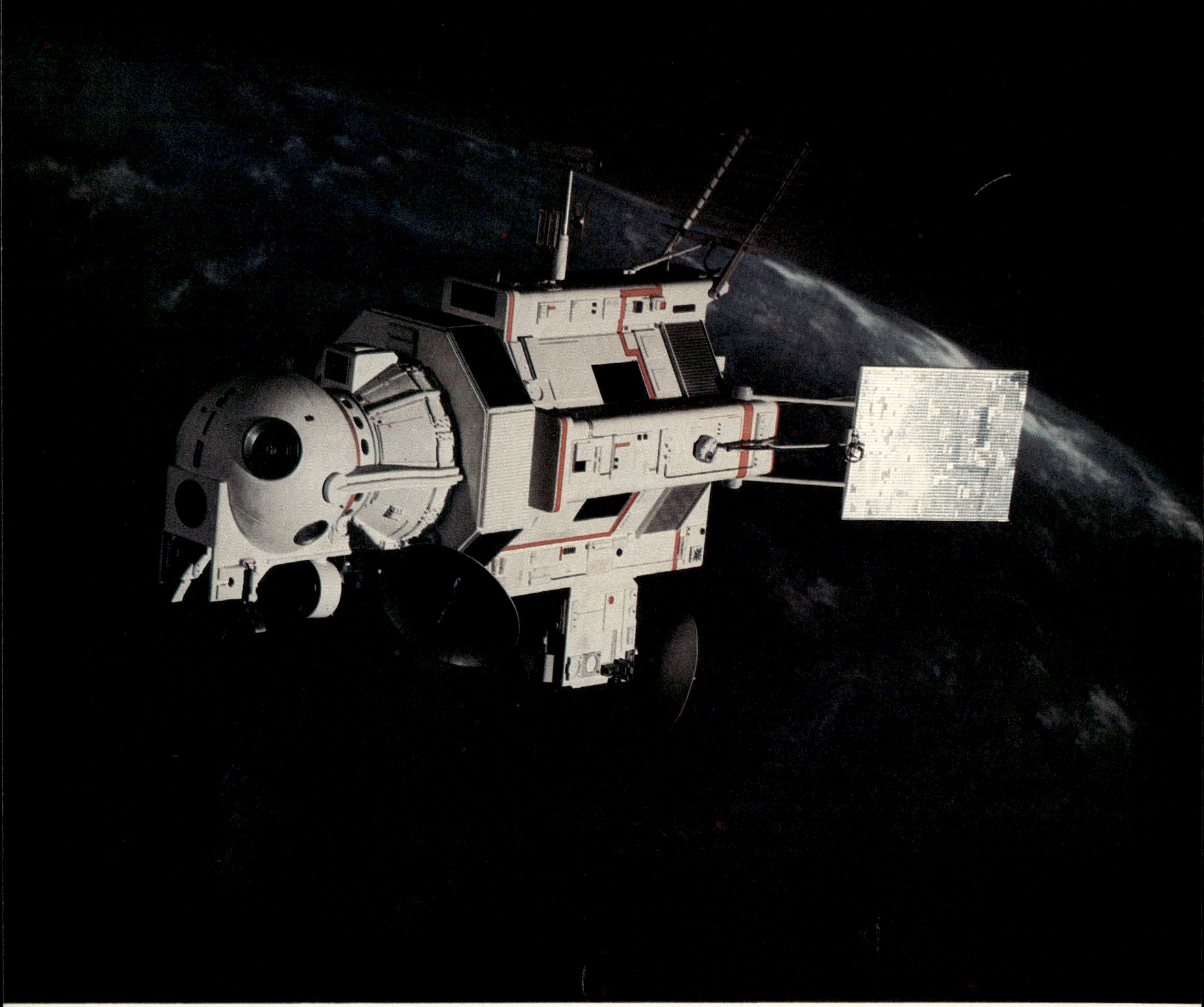

"S.I.D."—The Space Intruder Detector—orbits the Earth continually scanning the vastness of outer space for the dreaded alien enemy. S.I.D. reports both via videoteletype and by voice (as provided by BBC newscaster Mel Oxley).

ROBOTS ON TV

1940—1959

CAPTAIN VIDEO AND HIS VIDEO RANGERS

Dumont (defunct network), 1949-56, B&W, 30-min. episodes.

Captain Video was "an electronic wizard, master of Time and Space, and Guardian of the Safety of the World." The role was played from 1949 to 1950 by Richard Coogan; thereafter Al Hodge starred as the Captain. Ironically, what must at the time have seemed like a stroke of extreme good fortune proved to be Hodge's undoing: after the show ended in 1956, Hodge was typecast as Captain Video much as George Reeves would be typecast as Superman. Neither actor ever really worked again; Reeves committed suicide in 1959, and Al Hodge died early in '79 in a small hotel room in New York City, a victim of poverty and alcohol. However, in the early 50s Hodge cut quite a dashing figure as the Captain, even if his costume of surplus military uniforms (Eisenhower short jacket, baggy trousers, and laced-up combat boots) did make him look rather like the Chief Pilot of a seedy charter airline. His secret headquarters in the year 2254 A.D. were located deep inside a mountain peak. Assisted by the Video Ranger (Don Hastings), he used his vast arsenal of scientific equipment to aid various world governments and fight those evil misguided souls who were a threat to world peace. Among his weapons were the Cosmic Vibrator, which killed people by vibrating them to a state of exhaustion; a Scillometer which enabled him to see through walls; the Static Beam which threw out lethal bolts of electricity, and the Atomic Rifle. A curious feature of the program was the Captain's habit of showing old Western films on his Viewscreens; this was explained away by the news that the stars of these films were his secret agents in the West, and he was merely keeping an eye on what they were up to. The Captain was opposed by the evil Dr. Pauli (Hal Conklin), whose chief weapon was a device that made all sound cease, making it possible for him to sneak around without making any noise. Another villain was Dr. Tobor (Dave Ballard), a robot with padded tin shoulder wings, large crescent-shaped hooks for hands, two protruding eyes, and what looked like a periscope rising out of the front of his tin head. For all his strange looks, Tobor was the first "robot regular" on TV. It was several decades before the robots' TV image recovered. In 1953 the program was re-titled *THE SECRET FILES OF CAPTAIN VIDEO*: in 1956, the Captain was reduced

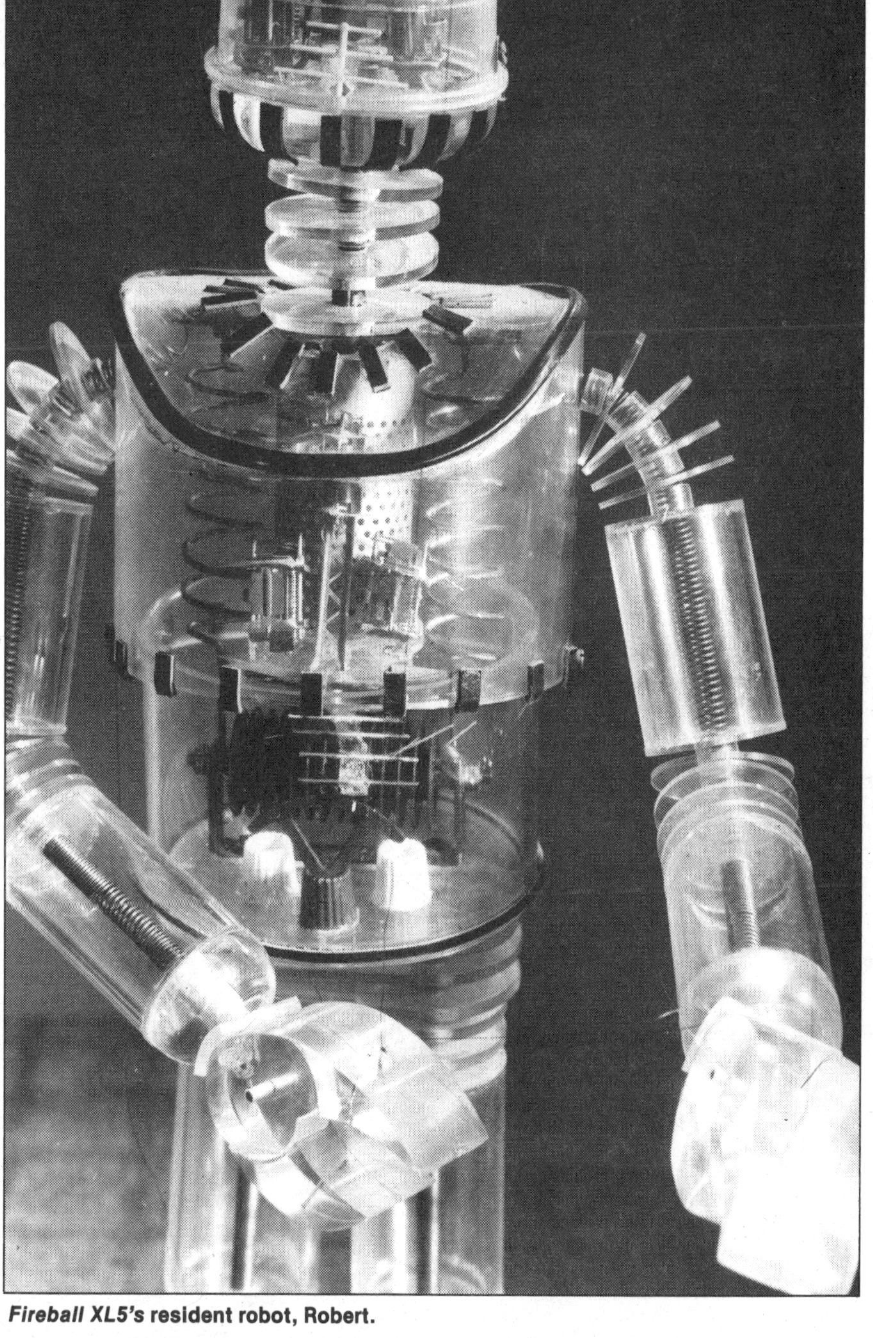

***Fireball XL5's* resident robot, Robert.**

to hosting a weekly presentation of old Betty Boop cartoons (voice: Mae Questel), and the show was re-titled *CAPTAIN VIDEO'S CARTOONS*.
Producer: James Caddigan. Director: Arnee Nocks, others. Writers: Maurice C. Brock, others.

SCIENCE FICTION THEATRE
NBC-TV, 1955-56, B & W, 30-min. episodes.
Truman Bradley hosted this SF anthology series, which always tried to present its bizarre stories in a very matter-of-fact manner. The show always opened with Bradley demonstrating the invention or scientific principle upon which the story to follow was based. These inventions included several robots and represented the first TV exposure of the current scientific state of the robotic arts. Unfortunately, production information about those specific episodes is not accessible.

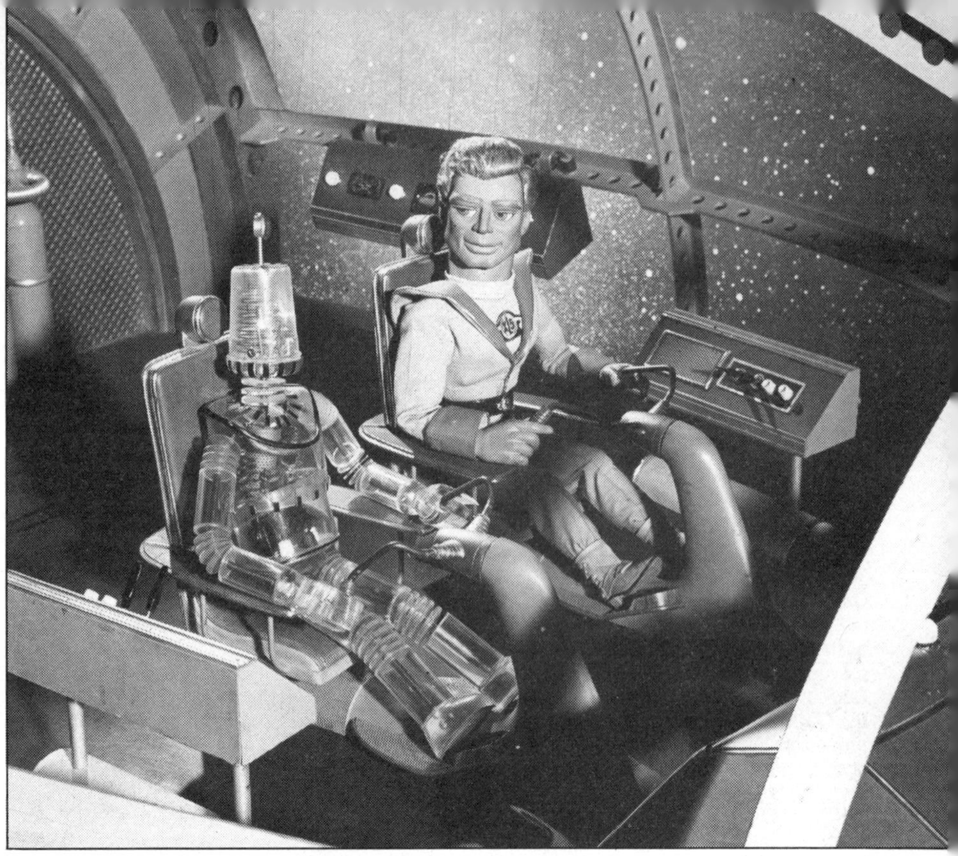

Robert the Robot helps Col. Steve Zodiac pilot *Fireball XL5*.

THE THIN MAN
NBC-TV, 1957-59, B&W, 30-min. episodes.
Based on characters created by novelist Dashiell Hammet, this detective series followed the adventures of Nick and Nora Charles (Peter Lawford, Phyllis Kirk). Nick, a former detective turned editor of a publishing house, spent much of his time solving the crimes his socialite wife kept stumbling across. Also featured were Nita Talbot, Jack Albertson, and Blanche Sweet.
Music: Pete Rugolo.
Robot Episode:
"The Robot Client": Robby the Robot makes a guest appearance as the new invention of a noted scientist. Unfortunately for Robby, it looks as though he may have murdered someone during his first public demonstration; he's destined for the scrap heap unless Nick can track down the real murderer. Robby seems to be operating on some version of Asimov's "Three Laws of Robotics."

The Jetsons' "U-Rent-A-Robot-Maid," Rosey.

SUPERCAR
Syndicated (ITC Production), 1959, B&W, 39 30-min. episodes.
Gerry and Sylvia Anderson's first "Supermarionation" effort, Supercar was an incredible device created by Professor Popkiss (voice: George Murcell) which could fly, submerge beneath the sea, and even travel on land occasionally. Driven by test pilot Mike Mercury (voice: Graydon Gould), assisted by Dr. Beaker and Jimmy Gibson (voices: David Graham, Sylvia Anderson), Supercar roamed the world helping our heroes fight crime.
Writers: Hugh and Martin Woodhouse. Special Effects: Derek Meddings. Music: Barry Gray.
Robot Episode:
"The Lost City": While on an expedition to the South Pole, Mike, Dr. Beaker and Jimmy stumble across the secret underground headquarters of mad Professor Watkins, located under the Amazon River basin. Watkins, protected by robot guards, plans to fire a deadly missile at Washington. The missile is deflected from its course with the aid of Supercar, giving Dr. Beaker enough time to change its course by radio jamming: the missile plunges back toward its launch site, destroying the underground city of Professor Watkins.

1960s

FIREBALL XL-5
Syndicated (ITC Production), 1961, B&W, 39 30-min. episodes. (Seen on NBC-TV 1963-64)

Created by Gerry and Sylvia Anderson, this was their second "Supermarionation" series, a process using electronically-controlled fiberglass marionettes as performers. *Fireball XL-5* is the lead spacecraft of the World Space Patrol, commanded by Steve Zodiac (voice: Paul Maxwell), assisted by Professor Matthew Matic (voice: David Graham) and Dr. Venus (voice: Sylvia Anderson). Their job is to patrol Sector 25 of the Universe, beyond our solar system. The Automatic Pilot for the spaceship is Robert the Robot (voice: Gerry Anderson), one of the few robots to be a regular on a television series. Robert stood 22 inches high, was transparent with all his parts showing, and had a tendency to heat up and smoke when things didn't compute properly. Executive Producer: Gerry Anderson. Writers: Anthony Marriott, Alan Fennell, Dennis Spooner. Art Director: Reg Hill. Cameraman: John Read. Special Effects: Derek Meddings. Music: Barry Gray.

Robot Episodes:

"The Wings of Danger": Steve Zodiac is attacked by a large robot bird built by the lawless Subterrains on Planet 46 in Sector 35. He captures the robot as it is attempting to implant a radium capsule in his body. Converting the bird to his own control, Steve takes the deadly robot back to Planet 46 and uses it as a threat to force the Subterrains to obey the Universe Code of Laws.

"The Granatoid Tanks": The crew of *Fireball XL-5* is forced to battle the Granatoids, a robot army of evil intelligence that intends to take over a planet that Earth is about to colonize. The Granatoids are self-programming and think independently, and are each contained within a spacecraft that can move across the surface of the planet like an Earth-type military tank. The robots are finally repelled by a strand of pearls worn by a stowaway on board the *XL-5*, which by sheerest chance are made of Plyton, the one material which Granatoids just *hate*. . . .

"The Robot Freighter Mystery": A series of explosions over a period of a few months have destroyed several of the unmanned space freighters controlled by robots and computers to ferry supplies around the galaxy. Steve Zodiac suspects the Briggs Brothers, owners of a salvage company, are sabotaging the freighters, but he can find no proof. By threatening to blow the Briggs Brothers up with a time-bomb, he forces them to confess their guilt. (It is at this point that we begin to suspect that Steve Zodiac is a fascist.)

"Trial By Robot": The evil Professor Himber has brought the robots of four planets under his control, removing them to an area of Planet 82 that he calls Robotvia. Robert the Robot also falls under Himber's control, and the crew of the *XL-5* is captured and put to work in Robotvia repairing robots. Professor Matic and Steve Zodiac devise a method of taking control of the robots back from Himber, and turn the kidnapped robots loose against him; apparently they are programmed for revenge.

Astro Boy battles enemy forces.

Dr. Bob McDonald and his *Living Doll,* Rhonda.

THE JETSONS

ABC-TV 1962-64, CBS-TV 1965-66, NBC-TV 1966-67, CBS-TV 1969-70, NBC-TV 1971-75, Color, 30 min. episodes.

The animated adventures of a wacky space-age family in the 21st Century. Synthetic food, anti-gravity devices, and robots abound. George O'Hanlon, Penny Singleton, Janet Waldo and Daws Butler were the voices of the Jetsons. Also featured were Mel Blanc, Don Messick, Jean VanderPyl, Shepard Menkin, and Howard Morris.

Executive Producers-Directors: William Hanna, Joseph Barbera. Writers: Warren Foster, Mike Maltese, Harvey Bullock, Larry Markes, Tony Benedict. Music: Hoyt Curtin.

ASTRO BOY

Syndicated, 1963, Color, 30 min. episodes

This was the first of many SF/Fantasy cartoon shows developed in Japan for the American market. It takes place in the 21st century, shortly after Dr. Boynton's wife and son have been killed in an auto accident. To ease his loneliness, he uses his scientific knowledge to create an indestructible robot companion, Astro Boy. Together, they fight crime, space monsters, and mad scientists, thanks to Astro Boy's super-powers.

DOCTOR WHO

Syndicated (Time-Life/B.B.C.), 1963-present, B&W and Color.

This popular British show is the oldest continuing SF series in the world: for over 15 years it has been a mainstay of British television, presenting the adventures of the indestructible and indefatigable Dr. Who. The good doctor is one of the Time Lords of the Planet Gallifrey, a sophisticated race who discovered time travel several thousand years ago (or yesterday or tomorrow, if you look at it that way, relatively speaking). After several thousand years of traveling in their T.A.R.D.I.S. (Time and Relative Distance In Space) machines, the Time Lords decided to abandon active time travel and become mere observers. This was not enough for Dr. Who: stealing a T.A.R.D.I.S. from the repair shop, he set off to experience the Universe directly. Captured, he is temporarily exiled to 20th Century Earth, where he resolves to use his extraordinary powers to defend good from evil. Four actors have played Dr. Who over the years: William Hartnell, Patrick Troughton, Jon Pertwee, and Tom Baker. Far from ignoring this, the show has made a point of explaining that the Time Lords are capable of regenerating their own bodies when threatened by old age or illness; each change of actor has merely been another regeneration. Likewise, the T.A.R.D.I.S. has a chameleon device enabling it to change its appearance to blend into any environment; however, the device on Dr. Who's T.A.R.D.I.S. has been jammed for several years, causing it to always appear as a London Police phone booth. The T.A.R.D.I.S. is also dimensionally transcendental, being much larger on the inside than on the outside. Much like the budget for the series; a great deal is done with imagination. Most of the Dr. Who episodes have never been seen in the U.S.; the episodes currently being shown are from seasons 11-14 (1974-78), with Tom Baker as Dr. Who.

Producer: Graham Williams. Visual Effects: Mat Irvine. Sound Effects: Dick Mills. Music: Dudley Mills, Ron Granier.

Robot Episodes:

"Robot" (1974-75): Recovering from his latest body regeneration, Dr. Who and his assistants Sarah Jane Smith and Harry Sullivan (Elisabeth Sladen, Ian Marter) investigate the theft of top secret computer codes and the plans for a disintegrator gun. Dr. Who must get past a killer robot and the ray gun to recover the computer codes before dissident scientists use them to launch nuclear missiles and begin World War III.

"Genesis of the Daleks" (1974-75): This episode re-introduces Dr. Who's old adversaries, the Daleks, familiar to American audiences from two films about Dr. Who which starred Peter Cushing. Dr. Who, Sarah and Harry are sent by the Time Lords to the Planet Skaro, which lies in ruins from a thousand-year war between the Kaleds and the Thals. Davros, the leader of the Kaleds, is in the process of turning his people into Daleks, evil creatures in wheeled robot bodies, whose only purpose is conquest. Dr. Who and his friends have arrived back in time before the process was complete, and Dr. Who tries to stop Davros from carrying out his plan.

"The Revenge of the Cybermen" (1974-75): Dr. Who and his two assistants land on a space station which is being taken over by Cybermen: humanoid in form, they have replaced their limbs and organs with bionic implants, becoming ruthless, emotionless silver giants in the process. Dr. Who enlists the aid of the Cybermen's hereditary enemies, the Vogans, to defeat them.

"Terror of the Zygons" (1974-75): Dr. Who discovers that the Loch Ness monster is actually a cyborg created by the evil Zygons, who have been living on the bottom of the loch since their spaceship crashed on Earth centuries earlier. Now convinced that repair of their ship is impossible, they have decided to use the cyborg to conquer the Earth.

Doctor Who's companion Leela is captured by "The Robots of Death."

"The Android Invasion" (1975-76): Dr. Who lands his T.A.R.D.I.S. in a peaceful English village, and is disturbed to discover that all its inhabitants are android robots. Then he discovers he is not even on Earth, but in a mock village built by the evil Kraals, who plan to use the androids to conquer the Earth.

"The Robots of Death" (1976-77): Dr. Who and his new travelling companion, Leela (Louise Jameson), discover a murder which they believe was committed by the worker robots in a mine; wheeled robots with claw hands. Since this is a violation of the robots' basic programming, the other humans do not believe this theory and accuse Dr. Who and Leela of the murder. Only when all the robots go berserk and turn on the shocked humans is the truth revealed that someone has re-programmed the robots into murdering fiends.

"The Invisible Enemy" (1977-78): After being infected with a killer virus, Dr. Who cures himself by having microscopic copies made of himself and Leela; together, they enter his body and fight the organism. As a reward for stopping the spread of the disease, Dr. Who is given a robot dog named K-9.

"Underworld" (1977-78): Dr. Who and Leela battle the Oracle, a megalomaniacal robot computer which has seized the race-banks of the Minyan civilization.

In addition to the robots described in the episodes above, there are a number of robots that the doctor encountered in his previous incarnations:

The Yeti: These were robots resembling the Abominable Snowmen of Tibet, created by an evil alien intelligence. Dr. Who was forced to fight them in the London Underground. It is thought possible that several escaped to New York and are responsible for much of the subway crime here; I myself am certain I once encountered one at 110th St. and Lenox Avenue.

The Autons: Robot plastic dummies; some were crude and unfinished, others were completely accurate replicas of human beings. They were created by the evil, formless Nestenes, whose sinister affinity for plastic could also turn plastic flowers and telephone cords into ruthless killers.
The Mechanoids: These were the globe-like robot inhabitants of the jungle planet Mechanus. The doctor barely survived a battle between the Mechanoids and their flame-thrower guns and some Daleks.
The Boss (A Super-Computer): This ruthless robot computer had taken over the operation of a giant chemical company. Its disregard of the environment caused a Welsh mine to become so polluted that a new disease developed: The Green Death, spread by giant maggots. New Yorkers are familiar with this one.

The Cybermen are bent on revenge against the televised *Doctor Who*.

MY LIVING DOLL

CBS-TV, 1964-65, 30-min. episodes.
Julie Newmar starred as Rhonda the Robot in this short-lived comedy series. Rhonda was the end result of a secret government space project, AF709. Though the government had done wonders with her physical development, she was sent to psychiatrist Dr. Bob McDonald (Bob Cummings) so that he might guide her emotional and mental development. She was capable of free thought, and several of the continuing characters hoped free love might have slipped in somewhere, too. The expected and usual comic complications followed as McDonald had to explain her presence in his house, keep her true identity secret, and keep her out of trouble. Also featured were Doris Dowling, Jack Mullaney, and Henry Beckman.

THE EIGHTH MAN

Syndicated, 1965, Color, 30-min. episodes.
Agent Peter Brady has been killed trying to catch arch-criminal Saucer Lip, who works for a mysterious society known as Intercrime. Professor Genius puts Brady's Life Force into a robot known as Tobor, the Eighth Man, which enables Brady to go on fighting crime in this Japanese cartoon series.

LOST IN SPACE

CBS-TV, 1965-68, B&W (season 1) and Color (seasons 2 & 3), 60-min. episodes.
The Robinson Family (Guy Williams as Professor John Robinson, June Lockhart as his wife June, Billy Mumy as Will, Marta Kristen as Judy, and Angela Cartwright as Penny) are traveling through space in the *Jupiter 2*, looking for new worlds to colonize as part of a government project. Along for the ride are Captain Don West (Mark Goddard) as co-pilot, the villainous Dr. Zachary Smith (Jonathan Harris), and a robot (Bob May; voice: Dick Tufeld). Dr. Smith's interference causes the ship to become hopelessly lost; the Robinsons spent the next three seasons seeking a way back to Earth while becoming involved with various alien cultures. The robot in *Lost In Space* was certainly the forerunner of STAR WARS' R2-D2 and C-3PO in terms of dramatic function. Though originally seen as an evil, independent machine in the first episode, he was quickly changed to a protective guardian. This change in character made the robot more human-like than most of the others. He was never given a proper name but constantly referred to as "robot." Dr. Smith also underwent a personality change: though still the mysterious agent of a foreign power, he became less evil and more of a buffoon as the series continued.
Creator-Producer: Irwin Allen. Special Effects: L B. Abbott, Howard Lydecker. Music: Lionel Newman, John Williams, Herman Stein.
Robot Episodes:
"The War of the Robots" (1965-66): Robby the Robot appears as a defunct robot created by an advanced civilization. Will restores it to working order, whereupon it decides to enslave the Earth people and destroy their own robot.
Director: Sobey Martin. Teleplay: Barney Slater.
"His Majesty Smith" (1965-66): An android robot society makes Dr. Smith its king; actually they have chosen him for his uselessness, and plan to end his reign with a human sacrifice.
Director: Harry Harris. Teleplay: Carey Wilber.
Guest Cast: Liam Sullivan, Kevin Hagen.
"The Ghost Planet" (1966-67): Smith sets the *Jupiter 2* down on an alien world populated by Cyborgs who decide to enslave the Earth people for manual labor. They are finally rescued by their own robot, the Uncle Tom of the robot world.
Director: Jerry Juran. Teleplay: Peter Packer.
Guest Cast: Sue England, Michael Fox.
"The Android Machine" (1966-67): The Robinsons encounter an emotionless android robot and teach her the meaning of love.
Director: Don Richardson. Teleplay: Rob and Wanda Duncan.
Guest Cast: Dee Hartford, Fritz Feld.
"The Dream Monster" (1966-67): A scientist with another emotionless android robot; this one tries to drain the Robinsons of all their emotions for transfer to his creation.
Director: Don Richardson. Teleplay: Peter Packer.
Guest Cast: John Abbott, Dawson Palmer, Harry Monty, Frank Delfino.

"Revolt of the Androids" (1966-67): Smith thinks he can transform an android robot into the most powerful creature in the Universe. He's right, but then what do you do with it?
Director: Don Richardson. Teleplay: Bob and Wanda Duncan.
Guest Cast: Dee Hartford, Don Matheson, Dawson Palmer.
"The Phantom Family" (1966-67): A scientist from Outer Space (Alan Hewitt) creates android duplicates of all the Earth people, then forces Will to instruct them in proper behavior.
Director: Ezra Stone. Teleplay: Peter Packer.
"The Mechanical Men" (1966-67): A horde of tiny robots cause the personalities of Dr. Smith and the Robinson robot to be exchanged, hoping to create the perfect leader; they are sadly disappointed in the result. The tiny robots were Remco plastic model kits of the big robot.
Director: Jerry Juran. Teleplay: Barney Slater.
"The Wreck of the Robot" (1966-67): The Robinson robot becomes the blueprint for a terrible machine with which evil aliens hope to conquer the Earth.
Director: Jerry Juran. Teleplay: Barney Slater.
Guest Cast: Jim Mills.
"The Condemned of Space" (1967-68): The *Jupiter 2* stops on a satellite to refuel and the Robinsons find themselves in the middle of a computerized space prison with Robby the Robot as a guard.
Director: Jerry Juran. Teleplay: Peter Packer.

Dr. Smith teaches the Robot the finer things in life in *Lost in Space*.

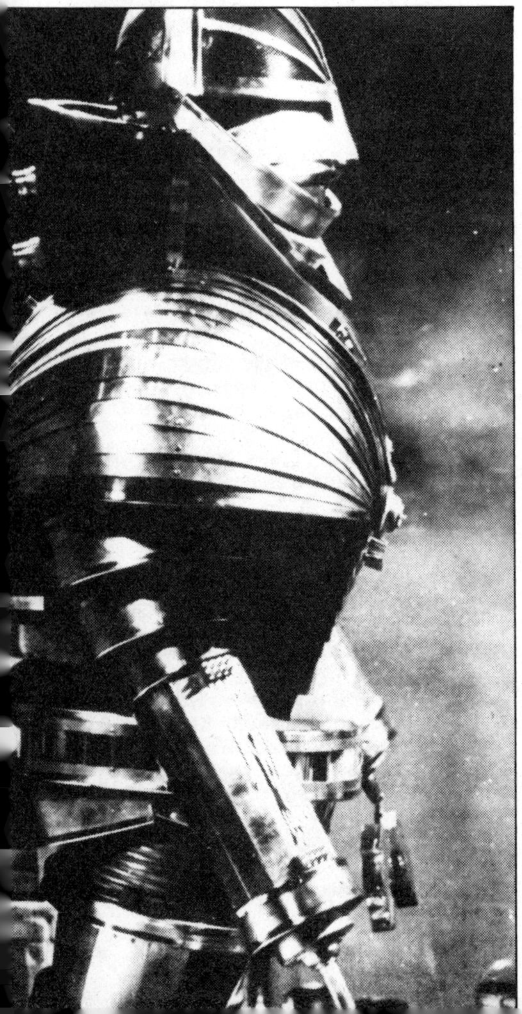

"The Robot" has been reprogrammed for murder in *Doctor Who*.

Guest Cast: Marcel Hillaire.
"Kidnapped In Space" (1967-68): A gang of androids kidnap the Robinsons, then force their robot to turn surgeon and operate on the malfunctioning android leader.
Director: Don Richardson. Teleplay: Robert Hamner.
Guest Cast: Grant Sullivan, Carol Williams, Joey Russo.
"The Space Destructors" (1967-68): Smith discovers a machine that makes androids, and dreams of building an army to conquer the Universe.
Director: Don Richardson. Teleplay: Robert Hamner.
Guest Cast: Tommy Farrell.
"Deadliest of the Species" (1967-68): The Robinson robot falls under the influence of a female robot programmed for evil. The robot has a lot of fun; the Robinsons do not.
Director: Sobey Martin. Teleplay: Robert Hamner.
Guest Cast: Ronald Gans, Lyle Waggoner.
"Junkyard of Space" (1967-68): The android owner of a junk planet absconds with the memory banks of the Robinsons robot, thanks to the nefarious Dr. Smith.
Director: Ezra Stone. Teleplay: Barney Slater.
Guest Cast: Marcel Hillaire.
"Hunter's Moon" (1967-68): Trapped on a strange planet, John Robinson is hunted as game by a humanoid named Megazor, with a robot as referee for the grim contest.
Director: J. Richardson. Teleplay: Jack Turley.
Guest Cast: Vincent Beck.

GET SMART

NBC-TV 1965-69, CBS-TV 1969-70, B&W and Color, 30-min. episodes.
Don Adams starred as C.O.N.T.R.O.L. agent Maxwell Smart in this spy-spoof comedy series. The agents of C.O.N.T.R.O.L. were dedicated to combatting the nefarious plans of K.A.O.S., an international organization of evil. Max was helped in his efforts by his Chief (Edward Platt) and Agent 99 (Barbara Feldon), and by Hymie, a C.O.N.T.R.O.L. robot (Dick Gautier). Hymie was self-programming and capable of free thought, though on occasion it wasn't too clear what the thoughts were. Also featured were Bernie Kopel, Victor French, and William Schallert.
Creators: Mel Brooks, Buck Henry. Producer: Leonard Stern. Music: Irving Szathmary.

GIGANTOR

Syndicated, 1966, Color.
This Japanese animated series featured the adventures of Gigantor, "The World's Mightiest Robot." Gigantor was controlled by 12-year old Jimmy Sparks, who used him to help Police Inspector Blooper fight intergalactic evil.
Music: Lou Singer.

THUNDERBIRDS

Syndicated (ITC Production), 1966, Color, 32 60-min. episodes.
Created by Gerry and Sylvia Anderson, this "Supermarionation" series used elec-

An alien robot killer-satellite hides behind space junk waiting to prey on unsuspecting spacecraft in *UFO*.

tronic marionettes as the performers. It told of the adventures of The Thunderbirds, members of an International Rescue team, whose home base was an ultra-scientific complex located on a hidden island. A secondary base was Thunderbird 5, an orbiting space station, and transportation between the two was provided by Thunderbird 3.

The bionic bigfoot of *Six Million Dollar Man*.

Executive producer: Gerry Anderson. Associate Producer: Reg Hill. Art Director: Bob Bell. Character Visualizer: Sylvia Anderson. Cameraman: John Read. Special Effects: Derek Meddings. Music: Barry Gray.

Robot Episodes:

"Sun Probe": A super-spaceship whose mission was to send a probe close to a solar flare to collect particles of matter has been caught in the Sun's gravitational pull and is about to be incinerated. Alan and Scott Tracy (voices: Matt Zimmerman, Shane Rimmer) set off to rescue the three astronauts aboard in Thunderbird 3. They succeed in putting the probe-ship back on course for Earth, but their own ship is itself now caught in the Sun's grip. A desperate calculation is needed to save them, but the only computer available is Braman, a malfunctioning robot invented by Brains (voice: David Graham), another member of the Thunderbirds team. Brains succeeds in repairing Braman, and the robot is able to correctly calculate a way to bring them back safely.
Director: Alan Patillo. Teleplay: Alan Fennell.

"30 Minutes After Noon": A gang of crooks attempt to blow up England's plutonium supplies using explosive bracelets set to go off at 12:30 P.M. A British agent trying to foil the plot is trapped inside the locked plutonium vault with the deadly bracelets by a team of non-self-programming robot guards. Using a Lasar Beam Director, the Thunderbirds cut through the vault doors, deactivate the robots, and dispose of the explosives before they blow up the plutonium.
Director: David Elliott. Teleplay: Alan Fennell.

STAR TREK

NBC-TV, 1966-69, Color, 78 60-min. episodes.

Television's first adult SF series with continuing characters, *Star Trek* told of the adventures of the Starship *Enterprise*, on a five-year mission of exploration, seeking out new life-forms and civilizations. Captain James T. Kirk (William Shatner) and his crew succeeded admirably in discovering all sorts of new life forms; what they failed to do was find enough viewers to keep the show on the air: it was cancelled after three seasons. Since then, however, the show has found a second life through syndication to local stations. Interest in *Star Trek* has remained at such a high level for so long that it will soon reappear from the Sargasso Sea of Old Television Shows as a major motion picture featuring all the old cast members (Leonard Nimoy as Vulcan Science Officer Spock, James Doohan as Engineering Officer Montgomery Scott, Walter Koenig as Lt. Chekov, DeForest Kelley as Dr. McCoy, Majel Barret as Christine Chapel, Nichelle Nichols as Lt. Uhura, and George Takei as Lt. Sulu) plus several new faces. The way things are going, the *Enterprise* may yet complete its five-year mission.
Creator and Executive Producer: Gene Roddenberry. Producers: Gene Roddenberry, Gene Coon, John Meredyth Lucas; also Fred Freiberger (third season). Story Consultants: D. C. Fontana, Steven Carabatsos. Directors of Photography: Ernest Haller, Jerry Finnerman. Special Effects: Howard Anderson Company, Westheimer Company, Film Effects of Hollywood, Jim Rugg. Makeup: Fred Phillips. Music: Alexander Courage, Gerald Fried. Costumes: Bill Theiss.

Robot Episodes:
"What Are Little Girls Made Of?" (1966-67): In this case, they are made of plastic and transistors, for they are android robots created by Nurse Chapel's fiance, Dr. Corby (Michael Strong). Aided by a giant android guard (Ted Cassidy as Ruk), Corby hopes to use his creations to take over the *Enterprise*. Before putting an end to his scheme, the crew learns that Corby is not the man they thought he was, in more ways than one: he, too, is an android.
Director: James Goldstone. Teleplay: Robert Bloch.
Guest Cast: Sherry Jackson, Harry Basch.
"The Changling" (1967-68): The *Enterprise* encounters Nomad, an ancient robot space probe which was damaged long ago in a collision with an alien robot probe capable of self-repair and maintenance; the two probes have combined into a super-powerful machine programmed to destroy all life-forms deemed imperfect. When it is brought on board the *Enterprise*, Kirk realizes he must find a way to deactivate this electronic fox in the hen-house before it discovers how surrounded by imperfection it is.
Director: Marc Daniels. Teleplay: John M. Lucas.
Guest Cast: Vic Perrin (Voice of Nomad), Barbara Gates, Arnold Lessing.
"The Doomsday Machine" (1967-68): The *Enterprise* rescues Commodore Decker (William Windom) from his disabled starship, the only survivor of an attack by a gigantic robot machine which devours whole planets for fuel to sustain itself. The Earth lies in its path, and Kirk must find a way to destroy this "doomsday machine" while dealing with Decker, who is obsessed with the fear that he failed earlier when a command decision was expected of him.
Director: Marc Daniels. Teleplay: Norman Spinrad.
Guest Cast: John Copage, Richard Compton, Elizabeth Rogers.
"I, Mudd" (1967-68): The *Enterprise* is hijacked by an android robot and taken to a planet inhabited by 2,000 robots who have made the notorious Harry Mudd (Roger C. Carmel) their ruler. Or so he thinks; Mudd is actually a captive, and the androids have brought the *Enterprise* crew to the planet merely to have more specimens to study. The crew devises a plan to overcome the powerful robots by shorting out their logic circuits through illogical action.
Director: Marc Daniels. Teleplay: Stephen Kandel.
Guest Cast: Kay Elliott, Michael Zaslow, Richard Tatro, Mike Howden.
"The Ultimate Computer" (1967-68): Complete control of the *Enterprise* has been turned over to a robot computer developed by Dr. Daystrom (William Marshall), as an experiment to see if human starship crews are obsolete. The computer is programmed with Dr. Daystrom's personality; unfortunately, he is mentally ill: the computer declares war on all other Federation ships, and will stop at nothing to keep control of the *Enterprise*.
Director: John Meredyth. Teleplay: D. C. Fontana, from a story by Lawrence N. Wolfe.
Guest Cast: Barry Russo, Sean Morgan.

1970s

ROD SERLING'S NIGHT GALLERY

NBC-TV, 1970-72, Color, 90 plays, 60-min. episodes (seasons 1 & 2), 30-min. episodes (season 3).
An occult anthology series created by Rod Serling, the series featured a gallery of strange paintings which Serling would use as an introduction to the evening's festivities. Sadly, Serling's connection with the show was pretty much limited to being the host, since he had signed away all creative control. Things reached such a state that he at one point tried to have his name removed from the credits, but contractual obligations prevented that. Following the show's cancellation he retired from television and became a college professor, a career which ended with his death in 1975.
Producer: Jack Laird. Makeup: Bud Westmore, John Chambers. Music: Gil Melle.
Robot Episodes:
"The Doll" (1970): A man returns to England after several years in India and discovers his niece playing with a doll that seems to be alive.
Director: Rudi Dorn. Teleplay: Rod Serling.
Guest Cast: Shani Wallis, John Williams, Henry Silva.
"You Can't Get Help Like That Anymore" (1971): The head of the Robot-Aids Agency warns a couple against abusing the robot help.
Director: Jeff Corey. Teleplay: Rod Serling.
Guest Cast: Broderick Crawford, Cloris Leachman, Lana Wood.

COLUMBO

NBC-TV, 1971-77, Color, 90-min. episodes.
This detective series featured a rumpled raincoat that solved crimes; or so it sometimes seemed. Peter Falk starred as Lt. Columbo, an ace detective disguised as an unmade bed, whose brilliant mind ticked away behind a mask of inept bumbling.
Creator: Richard Levinson. Executive Producer: William Link. Producer: Everett Chambers.
Robot Episode:
"Mind Over Mayhem": Robby the Robot made a guest appearance as the computer-like creation of a scientist. Sadly, his only function seemed to be to stand around and compute.

***The Six Million Dollar Man* uses his bionic limbs.**

The computerized *Future Cop* and his human partner.

THE EVIL TOUCH

Syndicated, 1973, Color, 30-min. episodes.
Anthony Quayle starred as the host of this Canadian anthology series. Each episode dealt with the plight of different people, all possessing the deadly seed of evil. One episode dealt with a politician who sent an android robot back in time to prevent the parents of his opponent from meeting, thus preventing the opponent's birth and eliminating him. Unfortunately, the android made a slight error, and prevented the meeting of the corrupt politician's parents instead.
Music: Laurie Lewis.

SPACE GIANTS

Syndicated, 1969, Color, 30-min. episodes.
This Japanese live-action series featured a whole troupe of giant marauding monsters out to ravage the Earth, led by Rodaka, an evil alien. Protecting the Earth was Goldar, a fifty-foot tall golden robot.

FRANKENSTEIN JR. AND THE IMPOSSIBLES

CBS-TV 1966-68, NBC-TV 1976-77, Color, 30-min. episodes.
The animated exploits of Frankenstein Jr., a thirty-foot tall robot with the voice of Ted Cassidy, as he battled evil. He was assisted in this endeavor by Coil Man, Fluid Man and Multi Man (voices: Hal Smith, Paul Frees, Don Messick), three U.S. government agents who pose as a rock-and-roll group: The Impossibles.
Executive Producers: William Hanna, Joseph Barbera. Director: Charles A. Nichols. Music: Hoyt Curtin.

LAND OF THE GIANTS

ABC-TV, 1968-70, Color, 60-min. episodes.
In 1983, while on a routine sub-orbital, 41-minute flight from Los Angeles to London, Flight 612 penetrates a solar turbulence area, and in a blinding flash of light passes into another dimension: the seven people on board discover they have come to Earth in a land where the inhabitants are 12 times their size. They are branded as freaks and fugitives: "The Little People." A reward is offered for their capture, and they are hunted by the giants as they try to find a way to return to their own world. The series starred Gary Conway, Don Matheson, Stefan Arngrim, Don Marshall, Deanna Lund, Heather Young, and Kurt Kaszner, as the fugitives from bigness.
Creator-Producer: Irwin Allen. Music: Lionel Newman.
Robot Episode:
"The Mechanical Man": Mark Wilson (Don Matheson) and Alexander Fitzhugh (Kurt Kasznar) are captured by Professor Gorn (Broderick Crawford), a giant scientist. Gorn has built a monstrous robot (James Daris), intending to use it as a psychological-warfare device, but the robot keeps malfunctioning. Gorn offers freedom to his two captives if they will repair it, so Mark climbs inside to do the necessary work. The robot finally turns on the evil Professor Gorn, allowing Mark and Fitzhugh to escape.
Director: Harry Harris. Teleplay: William L. Stuart.
Guest Cast: Stuart Margolin, William Chapman.

STAR TREK ANIMATED

NBC-TV, 1973-74, Color, 22 30-min. episodes.
This animated version of the original live-action series told of the further adventures of the Starship *Enterprise* and its crew, using the voices of William Shatner, Leonard Nimoy, DeForest Kelley, James Doohan, Majel Barrett, Nichelle Nichols, Walter Koenig, and George Takei in their original roles. While done in the usual Saturday-morning limited-animation, the stories were written by many of the original show's writers, placing it far beyond most of the competition in literary terms.
Producers: Norm Prescott, Lou Scheimer. Associate Director: Dorothy Fontana. Director: Hal Sutherland. Art Director: Don Ceristensen. Principal Character and Layout Designers: Bob Kline, Herb Hazelton. Background Color Director: Irv Kaplan. Editor: Joe Simon. Animation: Filmation Studios. Music and Effects: Horta-Mahana. Camera Supervision: R. W. Pope.
Robot Episodes:
"Once Upon A Planet": The *Enterprise* crew beams down to a computerized "shore leave" planet for rest and recreation, only to be confronted with threatening illusions which persist despite efforts to stop them. Kirk and Spock discover that the robot computer responsible for the illusions has come to resent its role of servant; Spock must logically convince the computer of the error of its ways before it kills them.
Writer: Len Jenson, Chuck Menville.
"Practical Joker": Attacked by two Romulan ships, the *Enterprise* takes refuge in a mysterious cloud, which causes the ship's computer to develop a peculiar sense of humor. The robot computer takes over the ship, playing practical jokes on the crew and then sending the ship hurling back toward the Romulan attackers.
Writer: Chuck Menville.

THE SIX MILLION DOLLAR MAN

ABC-TV, 1973-77, Color, 96 60-min. episodes.
After barely surviving a jet crash, test pilot Steve Austin (Lee Majors) is put back together using bionic components to replace the parts of his body destroyed in

The Crew of *The Lost Saucer.*

the crash, thereby becoming a Cyborg: a combination of human and robot components. His two bionic legs enable him to run at 60 miles per hour; he also now possesses a bionic arm and a bionic eye. All of this spawned another spin-off series, *The Bionic Woman*, numerous jokes about the possibility of another series called *The Bionic Chicken*, and at least one porno movie about further bionic possibilities. After recovering, Steve goes to work for the secret government agency that paid the $6,000,000 bill for his surgery, supervised by government agent Oscar Goldman (Richard Anderson). His bionic capabilities make him more than a match for his adversaries: thugs, spies, and the occasional robot.
Executive Producer: Harve Bennett.
Music: Oliver Nelson.

Robot Episodes:

"Day of the Robot": Steve must battle a robot (John Saxon) programmed to kill in order to prevent spies from stealing the secret of a secret missile guidance system.
Director: Les Martinson. Teleplay: Harold Livington, Del Reisman, Lionel E. Siegel.
Guest Cast: Henry Jones, Lloyd Bochner, Charles W. Bateman, Noah Keen.

"Run, Steve, Run": Steve is kidnapped by a crime syndicate which wants to use him as a model to build a team of bionic robots to rob Fort Knox.

Brian the Brain greets the personnel of Moonbase Alpha.

Zamara searches for "One Moment of Humanity."

Director: Jerry Johnson. Teleplay: Lionel E. Siegel.
Guest Cast: Noah Beery, Henry Jones, Mike Henry, George Murdock, Melissa Greene, Bill Conklin.

"The Seven Million Dollar Man": Another bionic man (Monte Markham) has been created who is more powerful than Steve Austin in physical strength; unfortunately, he is mentally unstable, and runs dangerously amok until subdued by Steve.
Director: Richard Moder. Teleplay: Peter Alan Fields. Guest Cast: Alan Oppenheimer, Maggie Sullivan, Fred Lerner, Marshall Reed.

"Return of the Robot Maker": A robot double of Oscar Goldman is used in a plot to destroy Steve Austin and steal the secret formula for a new energy source.
Director: Phil Bondelli. Teleplay: Mark Frost.
Guest Cast: Henry Jones, Ben Hammer, Troy Melton, Judd Laurance, Iris Edwards, Sarah Simmons.

"The Bionic Woman": Steve Austin's fiancee, Jaime Sommers (Lindsay Wagner), is critically injured in a skydiving accident. The same scientific techniques that restored Steve's smashed body after his plane crash are used to make Jaime the world's first bionic woman. She insists on using her new powers to help Steve break up an international counterfeiting ring.
Director: Richard Moder. Teleplay: Kenneth Johnson.
Guest Cast: Alan Oppenheimer, Malachi Throne, Martha Scott, Ford Rainey, Harry Hickox, Paul Carr, Elisabeth Brooks, Margaret Impert, Richard Jannone, Sidney Clute, Walt Conley.

"The Return of the Bionic Woman": Steve is shocked to learn that his fiancee, Jaime Sommers (Lindsay Wagner), who supposedly died when her body rejected her bionic implants, is alive; unfortunately, she is also suffering from amnesia. Steve attempts to restore Jaime's memory by taking her back to their hometown, but to no avail. Finally, he resolves to take her along on a dangerous mission involving a gang of terrorists, hoping that the challenge may unblock her memory of her past life and her love for him.
Director: Richard Moder. Teleplay: Kenneth Johnson.
Guest Cast: Martin E. Brooks, Richard Lenz, Ford Rainey, George Keymas.

"The Bionic Criminal": Barney Miller (Monte Markham), the second bionic man seen previously in the episode "The Seven Million Dollar Man," is reactivated and proceeds to use his super-powers for criminal purposes. Once again he is opposed by Steve Austin in a battle of bionic men.
Director: Leslie Martinson. Teleplay: Peter Allan Fields, Ron Bishop, Richard Carr.
Guest Cast: Alan Oppenheimer, Maggie Sullivan, John Milford, Donald Moffat.

"Death Probe": A giant robot space probe launched toward Venus by the Russians accidentally lands in Wyoming, where it begins destroying everything in sight. Steve Austin joins a team of American and Russian scientists who are attempting to stop the robot juggernaut before it wipes out a small town.
Director: Richard Moder. Teleplay: Steven E. deSouza.
Guest Cast: Martin E. Brooks, Nehemiah Persoff, Jane Merrow, Austin Stoker, John de Lancie, Phillip Pine, Beverly Garland, Walter Brooke, Don Dubbins, Ryan MacDonald, Ross Elliott, Bill Fletcher.

"Return of Deathprobe": Steve Austin battles a robot space probe programmed to destroy a city, but seems unable to stop the rampaging monster.
Director: Tom Connors. Teleplay: Howard Dimsdale.
Guest Cast: Martin E. Brooks, Than Wyenn, Ken Swofford, David Sheiner, Robert Lussier, Jimmy Joyce, John DeRose.

"Date With Danger": A robot computer gone wrong decides to take over the world by controlling all the world's other computers. It starts by taking control of the computers which handle the nation's secrets and money. Steve Austin is put to work to outwit the rogue machine, at

which point it decides to eliminate him.
Director: Rod Holcomb. Teleplay: John Meredyth Lucas, Wilton Schiller.
Guest Cast: Martin E. Brooks, Robert Walker Jr., Elaine Giftos, Luke Askew, Hank Brandt, Paul Tully, Robert Hackman, Peter Mark Richman, Noah Keen.

THE QUESTOR TAPES

NBC-TV, 1974, Color, 100 minutes.

This made-for-TV movie starred Robert Foxworth as Questor, an android robot whose mission is to prevent mankind from destroying the Earth. Unfortunately, while super-powerful and super-intelligent, Questor has been incompletely programmed by his creator, Dr. Vaslovik (Lew Ayres); the tape containing information on human emotions, drives and reasoning has been left out. After two attempts by scientific teams to reprogram him have failed, Questor seeks out a micro-electronics engineer (Mike Farrell) who provides the robot with the ability to cope with humanity and to seek out both Dr. Vaslovik and the secret of his creation. Made by Gene Roddenberry as the pilot film for a possible new series, the show failed to sell when Roddenberry refused to make several major changes required by the network.
Executive Producer: Gene Roddenberry. Director: Richard A. Colla. Screenplay: Gene Roddenberry and Gene L. Coon, from a story by Roddenberry. Art Director: Phil Barber. Special Photographic Effects: Albert Whitlock. Music: Gil Melle. Supporting Players: James Shigeta, Dana Wynter, Robert Douglas, Majel Barrett, Walter Koenig, Ellen Weston, John Vernon.

THE LOST SAUCER

ABC-TV, 1975-76, Color, 30-min. episodes. Incorporated into "The Krofft Supershow" 1976-77.

A comedy series about two androids (Ruth Buzzi and Jim Nabors) carting a young boy and his babysitter (Jarrod Johnson, Alice Playten) around in a flying saucer in an effort to get them back to Earth. Their other cargo is a dorse: a cross between a dog and a horse. You get the idea.
Executive Producer: Si Rose. Producers: Sid and Marty Krofft. Special Effects: Gordon Graff. Costumes: Jeremy Railton.

Holmes and Yoyo **demonstrate a robot's practical uses.**

SPACE: 1999

Syndicated (an I.T.C. Production), 1975-77, Color, 48 60-min. episodes.

By the year 1999, Earth is using the dark side of the Moon as a dump for nuclear waste materials. The entry of a wandering planet called Meta into the solar system causes the nuclear wastes to explode, blasting the Moon from its orbit and sending it hurtling off into the vastness of space. It evidently hurtles rather quickly, since in a remarkably short span of time the Moon and the inhabitants of Moonbase Alpha encounter numerous planets filled with unfriendly life forms. Martin Landau starred as Commander John Koenig, Alan Carter as Nick Tate and Barbara Bain as Dr. Helena Russell. Barry Morse was seen as Victor Bergman, but departed after the first season, as did Producer Sylvia Anderson. The second season brought a number of changes, notably an improvement in the scripts (at least in clarity) and several new characters: Catherine Schell as Maya and Tony Anholt as Tony Verdeshi. In spite of its faults, the show functioned very well on a purely visual level, with spectacular special effects and model work. Sadly, the show ran out of time while still making efforts to improve, being cancelled after the second season.
Executive Producer: Gerry Anderson. Producer: Sylvia Anderson (season 1); Fred Freiberger (season 2). Creators: Gerry and Sylvia Anderson. Special Effects: Brian Johnson. Costumes: Rudi Gernreich. Music: Barry Gray.

Robot Episodes:

"The Guardian of Piri" (1975-76): Lured to the planet Piri, the Alphans find a computerized mechanical paradise where all wishes and desires can be satisfied, courtesy of the Guardian, a giant robot computer. Koenig, seeing the living death implicit in such a purposeless life, resists the advances of the Servant of the Guardian (Catherine Schell, later to join the show as a regular as Maya) and fights to free his fellow Alphans from the Guardian's influence.
Director: Charles Crichton. Teleplay: Christopher Penfold.
"The Infernal Machine" (1975-76): Koenig and Dr. Russell enter an alien ship built by an old man (Leo McKern) which is a computerized robot, using his personality as its activating force. The old man dies, and the robot ship decides to keep Koenig and Russell as its human companions for the rest of their lives.
Director: David Tomblin. Teleplay: Anthony Terpiloff, Elizabeth Barrows.
"One Moment of Humanity" (1976-77): Dr. Russell and Tony Verdeshi are abducted by Zamara (Billie Whitelaw), the leader of a society of android robots. On Zamara's planet, the androids hope to use the two humans to learn the secret of emotions, so that they can become truly human.
Director: Charles Crichton. Teleplay: Tony Barwick.
"Brian the Brain" (1976-77): The Swift, a spacecraft from Earth, lands at Alpha piloted by a robot named Brian. At first Brian seems harmless, but then he kid-

naps Commander Koenig and Dr. Russell, and takes them to the planet where the human crewmen of his spaceship died. Maya and Tony Verdeshi plan a rescue mission, while Alpha is rocked by intense gravitational disburbances.
Director: Kevin Conner. Teleplay: Jack Ronder.
Guest Cast: Bernard Cribbins as Captain Michael.

HOLMES AND YOYO

ABC-TV, 1976, Color, 30-min. episodes.
This short-lived comedy series starred John Shuck as Gregory "Yoyo" Yoyonovich, a self-programmable police robot who never seemed to work very well. Richard B. Shull co-starred as Alexander Holmes, Yoyo's long-suffering partner in crime-fighting.
Creators: Jack Sher, Lee Hewitt. Executive Producer: Leonard B. stern. Producer: Arne Sultan.
Supporting Players: Bruce Kirby, Andrea Howard.

FUTURECOP

ABC-TV, 1976, Color, 5 60-min. episodes.

Rem makes some personnal repairs in *Logan's Run.*

The network's attempt to make the *Holmes And Yoyo* idea of an android policeman work seriously. Michael Shannon starred as the rookie cop who is really an android, with Ernest Borgnine as his unsuspecting partner. After the android is damaged, his human partner realizes the truth and proceeds to exploit the abilities of the robot for his own gain. Eventually, however, the two become friends, and the android sacrifices itself to save the human cop. John Àmos appeared as Officer Bill Bundy.
Creators: Anthony Wilson, Allen Epstein. Executive Producers: Anthony Wilson, Gary Damsker. Producer: Everett Chambers.

THE BIONIC WOMAN

ABC-TV, 1976-78, Color, 57 60-min. episodes.
After making several appearances as Steve Austin's bionically-repaired fiancee

***Jason of Star Command's* mini-robot, Wiki.**

on *The Six Million Dollar Man*, Jaime Sommers (Lindsay Wagner) was spun-off into her own series. Taking a cover job as a telephone book-ripping school teacher, she led a double life as a secret agent for the government's Office of Scientific Investigation, putting her bionic abilities to work fighting spies and criminals. Richard Anderson, who played Oscar Goldman, the head of OSI, in *The Six Million Dollar Man*, repeated his role in this series, shuttling back and forth between the two for the several years they were on simultaneously.
Executive Producer: Harve Bennett. Music Supervision: Hal Mooney.
Robot Episodes:
"Doomsday Is Tomorrow": A sophisticated robot computer has been built by scientist Elijah Cooper (Lew Ayres) to blackmail the world into lasting peace. The computer controls a Doomsday Bomb, and Jaime Sommers is thrown into a desperate duel with the computer for control of the device.
Writer-Director: Kenneth Johnson.
Guest Cast: Martin E. Brooks, Sam Chew, David Opatoshu, Ken O'Brian, James Hong, Guerin Barry, Stack Pierce.
"Kill Oscar": An embittered scientist (John Houseman) has created robot replicas of the OSI secretaries in order to abduct Oscar Goldman; Jaime and Steve Austin must join forces to save him from the madman and his fembots.
Director: Alan Crosland. Teleplay: Arthur Rowe.
Guest Cast: Martin E. Brooks, Janice Whitby, Jack Colvin, Corinne Michaels, Jennifer Darling, Jack L. Ging, John Dewey-Carter.
"The Bionic Dog": Yes, yes . . . Jaime discovers the existence of the world's first bionic canine, Maximillian, and has only five days to save it from certain death. The two cyborgs become fugitives in the Sierra Nevada mountains escaping from a raging forest fire.
Director: Barry Crane. Teleplay: James D. Parriott, from a story by Parriott and Harve Bennett.
Guest Cast: Martin E. Brooks, Ford Rainey, Taylor Lacher, Carlene Watkins, Dale Robinette, Lee Jones-De-Broux, Jack Garner, Jason Johnson, Al Hansen.
"Fembots In Las Vegas": The deranged son (Michael Burns) of a deranged scientist reactivates his father's army of female robots, using them to steal America's secret directed-energy ray weapon. After launching it into space, he demands Jaime, Oscar, and Dr. Rudy Wells (Martin E. Brooks) as ransom or he will turn the weapon on the Earth.
Director: Michael Preece. Teleplay: Arthur Rowe.
Guest Cast: James Olson, Jennifer Darling, Melinda Fee, Jeannie Wilson, Nancy Bleier, Lorna Sands, Lisa Moore, Alexander Courtney, Ted Schliesman.

SPACE ACADEMY

CBS-TV, 1977-present, Color, 30-min. episodes.
Located in a super-scientific complex built on an asteroid, the Academy is the training school for a group of young cadets (Pamelyn Ferdin, Rick Carrott, Ty Henderson, Maggie Cooper, Eric Greene, Brian Tochi). Their instructor is the seemingly ageless Commander Gampu (Jonathan Harris), who teaches his pupils the meaning of life through adventures on different worlds and through contact with different life-forms. He is aided in this task by Peepo, a small, wheeled robot who speaks with a constricted woman's voice and is noticeably lacking in courage. However, the importance of this little 'bot should not be underestimated: Peepo is the first (radio-controlled) robotic machine ever to be featured as a continuing character on a TV series.
Executive Producers: Lou Scheimer, Norm Prescott. Director of Photography: Alric Edens. Miniatures: Robert A. Maine, Chuck Comiskey, Paul Huston. Special Effects: John Frazier. Costumes: Bill Campbell. Art Directors: Ray Beal, William McAllister. Music: Yvette Blais, Jeff Michael.

PROJECT: U.F.O.

NBC-TV, 1977-78, Color, 60-min. episodes.
This Jack Webb series follows the adventures of two investigators for the U.S. Air Force's Project Blue Book as they try to explain mysterious sightings of unidentified flying objects around the country.

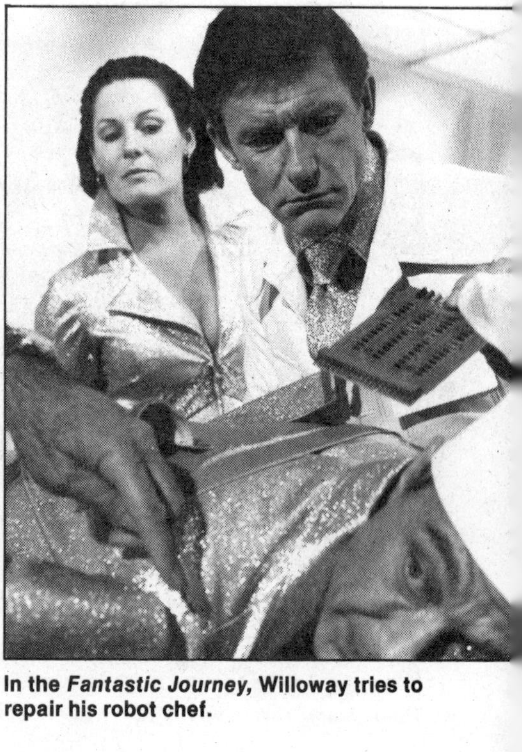

In the *Fantastic Journey*, Willoway tries to repair his robot chef.

7-Zark-7, the robot leader of G-Force in *Battle of the Planets*.

William Jordan starred as Major Jake Gatlin during the first season, but left the show to concentrate on film work and was replaced by Edward Winter as Captain Ben Ryan. Caskey Swaim was featured both seasons as Sergeant Harry Fitz. Like Webb's other shows, the stories in *Project U.F.O* were based on actual cases in official files. Realism was further advanced by producer William T. Coleman, himself an ex-Bluebook investigator.
Executive Producer: Jack Webb. Producer: Colonel William T. Coleman, U.S.A.F. (ret.). Special Effects: Don Weede, Wally Gentleman. Miniatures: Brick Price.
Robot Episodes:
"Sighting 4001: The Washington, D.C. Incident": Martha Carlyle (Frances Reid), a deeply religious farm woman, reports that an alien spacecraft landed in her front yard, and a robot got out to take soil samples. The robot communicated telepathically with Martha through a laser grid, telling her she had nothing to fear. Concurrent with this sighting, an Air Force pilot was killed in a crash while pursuing a UFO over Washington, at about the same time the UFO in Martha's front yard took off with its robot passenger.
Director: Richard Quine. Teleplay: Harold Jack Bloom.
Guest Cast: Anne Schebeen, John Findlatter, Len Wayland, Linwood McCarthy, William Wellman Jr., Tim Donnelly.
"Sighting 4010: The Waterford Incident": A boys military school is enveloped by a strange web-like substance that witnesses report was dropped from a UFO. Meanwhile, another witness to a sighting is afraid to talk about it for fear everyone will think he was either drunk or crazy, and he'll lose his job. He finally reports that while hunting in the woods he was chased by a robot which emerged from a triangular UFO.
Director: Dennis Donnelly. Teleplay: Michael Donovan.
Guest Cast: Craig Stevens, Dr. Joyce Brothers, Anthony Geary, Shane Sinutko, Winter Horton, Adam Gunn, Paul Napier.

ARK II

CBS-TV, 1976-77, Color, 30-min. episodes.

A live-action children's action-adventure series set in the year 2476 A.D., it concerned the efforts of a group of young people to re-establish civilization in a world destroyed by atomic war. It starred Terry Lester as Jonah, Jean Marie Hon as Ruth, and Jose Flores as Samuel. Together they faced the future, the dregs of society, and robots both good and bad, including Robby, who turned out to be both.
Executive Producer: Lou Scheimer, Norm Prescott. Producer: Richard Rosenbloom. Directors: Ted Post, Hollingsworth Morse.

THE FANTASTIC JOURNEY

NBC-TV, 1977, Color, 1 90-min. episode, 9 60-min. episodes.

While sailing through the Bermuda Triangle, a team of scientists pass through a strange green cloud and are shipwrecked on an uncharted land mass. Three of the survivors (Carl Franklin as Dr. Fred Winters, Scott Thomas as marine biologist Paul Jordan and Ike Eisenmann as Paul Jordan's son, Scott) encounter Varian (Jared Martin), a man from the 23rd Century who for some strange reason is trying to pass himself off as an Indian. Once exposed (as it were), Varian explains that his spacecraft has crashed and that he, too, is stranded on the land mass. He also explains that different time spheres exist simultaneously in this "vortex," and that by traveling Eastward one moves through the past, present and future. Varian, Dr. Winters, and the two Jordans all begin the journey, hoping to find a way back to their own time zones.
Executive Producer: Bruce Lansbury. Producer: Leonard Katzman. Story Consultants: D. C. Fontana, Calvin Clements, Jr. Music: Robert Prince.
Robot Episode:
"Beyond the Mountain": The time travelers have been joined by Liana the Atlantean (Katie Saylor), who is separated from them during a crimson storm. She finds a village of beautiful men and women who turn out to be android robots, led by Jonathan Willoway (Roddy McDowell), a scientist from the 1960s. Meanwhile, Varian, Dr. Winters and Scott Jordan are lost in a swamp, where they have met some green-skinned creatures who claim to be the rightful owners of

Willoway's androids, and who are determined to get them back.
Director: Irving Moore. Teleplay: Harold Livingston.
Guest Cast: John David Carson, Marj Dusay, Lester Fletcher, Frank Corsentino, Joseph Pellasorte, Ron Burke.

LOGAN'S RUN

CBS-TV, 1977-78, Color, 14 60-min. episodes.

A spin-off from the motion picture of the same name, *Logan's Run* starred Gregory Harrison as Logan and Heather Menzies as Jessica, two "runners" from their oppressive future world in which no one is permitted to live past the age of 30. Aided by Rem, a self-programming, self-servicing android robot (Donald Moffatt), Logan and Jessica must make their way across hostile terrain to Sanctuary, an area where they will be safe from the pursuing Sandmen. Logan, as a renegade Sandman himself, is an especially wanted "runner," and is hunted with vigor by his old friend, Francis (Randy Powell). What started as an intelligent commentary on a perverted system of societal organization soon sadly degenerated into yet one more variation of *The Fugitive*, and became nothing more than an extended chase melodrama. It is of interest in that one of its major continuing characters was a robot, and other androids and more machine-like robots were prominently seen.
Executive Producers: Ivan Goff, Ben Roberts. Producer: Leonard Katzman. Story Editor: D. C. Fontana. Production Designer: Mort Rabinowitz.
Robot Episodes:
"Logan's Run" (90-min. pilot episode): Using scenes from the MGM feature film, this episode sets up Logan's flight from the City of Domes with Jessica. With Logan's old friend and fellow Sandman Francis in pursuit, they come to a Mountain City. Two robots in the city, Draco (Keene Curtis) and Siri (Lina Raymond), imprison them so the robots will have someone human to serve. They are rescued by the android Rem (the robots' repairman) and the trio set off in a hovercraft to find Sanctuary.
Director: Robert Day. Teleplay: William F. Nolan, Saul David, Leonard Katzman.
Guest Cast: Ron Hajek, J. Gary Dontzig, Anthony de Longis.
"The Innocent": Escaping from Francis and the other Sandmen, Logan, Jessica and Rem come across a strange computerized world whose only inhabitants are a beautiful young woman and two robot companions. She falls in love with Logan and makes plans to dispose of Jessica, her rival for Logan's affections.
Director: Michael Preece. Teleplay: D. C. Fontana, Ray Brenner.
Guest Cast: Lisa Eilbacher, Lou Richards, Barney McFadden, Brian Kerwin, Gene Tyburn.
"Futurepast": Rem becomes romantically involved with Ariana (Mariette Hartley), another android, a condition made obvious because he literally sparks whenever he is near her. In order to fulfill her function of Dream Analysis she attaches analysis devices to Logan and Jessica, a dangerous procedure which can sometimes kill the subject.
Director: Michael O'Herlihy. Teleplay: Kathryn Michaelin Powers.
Guest Cast: Michael Sullivan, Janis Jamison, Joey Fontana, Ed Couppee.

JASON OF STAR COMMAND

***CBS-TV, 1978-present, Color, episodes of varying length incorporated as part of* Tarzan And The Super 7.**

A spin-off from *Space Academy*, Star Command is a secret section of the Academy, and Jason (Craig Littler) is one of its best agents. He is assisted in his fight against the evil Dragos (Sid Haig), commander of the Dragon Ship, by peculiar Dr. Parsafoot (Charlie Dell) and nubile Nicole (Susan O'Hanlin). Controlling their efforts against evil is Commander Canarvin (James Doohan). Robots play an important role in almost every episode: in addition to Peepo, the waist-high wheeled robot familiar from episodes of *Space Academy*, there is Wiki, Dr. Parsafoot's new invention. Wiki is the ultimate utility-belt accessory: standing about 6 inches high, Wiki is mobile on two mechanical feet, and can be used as a laser, an x-ray device, and a tiny rocket ship. Wiki also speaks, though in computer language that perhaps only R2-D2 could understand.
Executive Producer: Lou Scheimer, Norm Prescott. Producer-Director: Arthur Nadel. Art Director: Bill McAllister. Special Effects: Gordon Graff, Bruce Behan. Miniature Photography: Jim Veilleux, Diana Wooten. Stop-Motion Animation: Jim Aupperle, Stephen Czerkas.

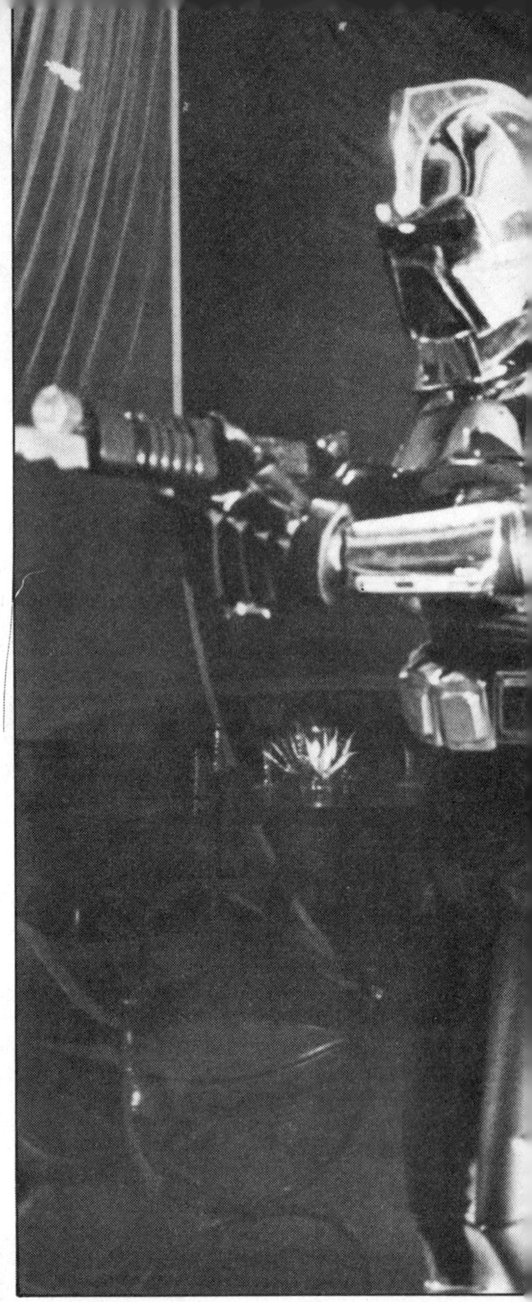

Below Left: Boxey and his pet Daggit, Muffit II.

BATTLESTAR GALACTICA

ABC-TV, 1978-79, Color, 60-min. episodes.

The most ambitious and expensive television series in the history of the medium, with absolutely magnificent special ef

Above: The Cylons advance on the colonists.

fects and miniature photography. Unfortunately, the scripts left much to be desired; the characters seem indistinguishable from one another: Starbuck smokes cigars but Apollo doesn't. Anyway, the story: After the sabotage of their 12-colony peace mission and the destruction of their home planets, the final remnants of human life in the galaxy join together in order to reach a world where humanity can safely flourish and rebuild itself: the distant and unknown thirteenth colony, Earth. Joining themselves into a rag-tag fleet of some 200 space vessels of all types and conditions, the pioneers are led by the *Galactica*, the only Battlestar which survived the treacherous ambush of the colonial peace mission. The *Galactica* is a ship of many purposes: Communications center, hub of the government, and the only source of protection against the pursuing Cylons, robot warriors determined to exterminate all human life. Lorne Greene stars as Adama, commander of the *Galactica* and leader of the immigrant fleet by virtue of his military and political rank. His son, Capt. Apollo (Richard Hatch), is the leader of the *Galactica*'s Fighter Squadron, assisted by his best friend, the fast-talking con-artist Lt. Starbuck (Dirk Benedict). Also featured are Herb Jefferson, Jr. as Boomer, Maren Jensen as Athena, Noah Hathaway as Boxey, and Terry Carter as Col. Tigh. The *Galactica* moves through the Universe, threatened by alien beings, natural disasters, and the ever-pursuing robot Cylons.

Executive Producer: Glen A. Larsen.
Supervising Producer: Leslie Stevens.
Producer-Special Effects: John Dykstra.
Costumes: Jean-Pierre Dorleac. Director of Photography: Ben Colman. Art Director: John E. Chilberg. Music: Stu Phillips, Glen Lanson.

Robot Episodes:
Strictly speaking, every episode involves robots, either the stainless-steel Cylons or the glass-domed Lucifer or even their awesome Imperious Leader, or Muffit, the robot canine daggit created by the Galactica's technicians as a pet and companion for Boxey (Noah Hathaway), Capt. Apollo's young son.

BATTLE OF THE PLANETS

Syndicated (Gallerie International Films), 1979, Color, 85 30-min. episodes.

Based on several full-length Japanese films, this animated series recounts the adventures of the superheroes of "G-Force" as they defend the Earth from alien invaders from the planet Spectra. The agents of G-Force are commanded by a robot leader, 7-Zark-7, and are aided by a robot dog, 1-Rover-1. They battle the evil forces of Spectra in their spaceship, *The Flying Phoenix*, which can alter its shape to that of its mythological namesake: a giant bird which could be re-born after making an ash of itself. The forces of Spectra are commanded by the evil Zoltar (voice: Key Luke), who regularly sends robot warriors against the Earth; so far, these have included robot turtles, robot squid, a robot praying mantis, and a thousand-eyed monster with a death ray in each eye.

Producer-Director: David Hanson.
Cast Voices: Alan Young, Casey Kasem, Janet Waldo, Ronnie Schell, Alan Dinehart.

ACKNOWLEDGEMENTS

Our grateful appreciation to the following people for supplying rare photographic material:

Bill Malone Robert Skotak Scot Holton
Jeff Sillifant Wade Williams

The information concerning the films and TV programs listed in this book was compiled from the following sources:

THE AMERICAN FILM INSTITUTE CATALOG OF MOTION PICTURES
Volume F2: Feature Films 1921-1930; Kenneth W. Munden, Executive Editor.
Volume F6: Feature Films 1961-1970; Richard P. Krafsur, Executive Editor.
The R.R. Bowker Company; New York.

THE BILLY ROSE THEATRE COLLECTION
New York Public Library at Lincoln Center, New York, N.Y.

THE BRITISH FILM CATALOG: 1895-1970
by Dennis Gifford.
McGraw-Hill Book Company; New York.

THE COMPLETE ENCYCLOPEDIA OF TELEVISION PROGRAMS
1947-1976
by Vincent Terrace.
A.S. Barnes and Company; New York.

FILMFACTS
Division of Cinema of the University of California; Los Angeles.

FROM JULES VERNE TO STAR TREK
by Jeff Rovin.
Drake Publishers, Inc.; New York.

THE GREAT SCIENCE FICTION PICTURES
by James Robert Parish and Michael R. Pitts.
The Scarecorw Press, Inc.; Metuchen, N.J.

HORROR AND SCIENCE FICTION FILMS
by Donald R. Willis.
The Scarecrow Press, Inc. Metuchen, N.J.

REFERENCE GUIDE TO FANTASTIC FILMS
by Walt Lee, assisted by Bill Warren.
Chelsea-Lee Books; Los Angeles.

ROBOTS, ROBOTS, ROBOTS
edited by Harry M. Geduld and Ronald Gottesman.
New York Graphic Society; Boston.

TV SEASON 74-75
TV SEASON 75-76
TV SEASON 76-77
compiled and edited by Nina David.
Oryx Press; Phoenix.

ABOUT THE AUTHOR

ROBERT M. HEFLEY is a professional actor based in New York City. Born in Savannah, Georgia, he attended the University of Florida, the Instituto Tecnologico de Monterrey in Mexico, and graduated from Florida State University with a BA in Theatre Arts. He served two years in the U.S. Army, assigned to the Special Services Entertainment Office in Bad Kreuznach, West Germany. His career began at Ivan Kivitt's Merry-Go-Round Playhouse in Miami, Florida. Since then he has acted in or directed more than 200 plays, with engagements at the Shady Lane Playhouse in Illinois, the Heritage Repertory Company in Virginia, the Barn Dinner Theatre Chain, the Carolina Regional Theatre, and others around the country. New York audiences have seen him in "Another Evening With Chekhov", "A Ritual of Absurdity" and playing the title role in "Iscariot." His previous work for STARLOG includes the "Television and Motion Picture Address Guides" in issues 5 & 6.